Rehabilitation Medicine and Thermography

Edited by

Mathew H.M. Lee, MD

Jeffrey M. Cohen, MD

Rehabilitation Medicine and Thermography

Published by Impress Publications, 28920 SW Meadows Loop, Wilson-ville, OR 97070 through Lulu Enterprises, Inc., 3131 RDU Center, Suite 210, Morrisville, NC 27560, www.lulu.com.

Copies of this book may be purchased online at www.lulu.com or from many online booksellers.

Library of Congress Control Number: 2008920834

ISBN: 978-0-6151-8721-1

Printed in the United States of America

Preface

In this book with its visual contents and varied topics, we hope to encourage our readers to explore the use of thermography in the clinical practice of chronic pain management for its diagnostic and prognostic role to relieve suffering!

From its diagnoses to therapeutic approaches to pain reduction, chronic pain is one of the most difficult conditions to manage.

Many approaches have been utilized to measure and quantitate pain. The technological advances in computerized thermography present us with an extraordinary tool.

Motor function, sensory patterns, coupled with thermographic patterns, we believe, will offer a huge diagnostic advantage and should become a standard triad in routine examinations.

Our earliest work in thermography was done at Goldwater Memorial Hospital studying low back and neck pains. This was followed by our seminal work on acupuncture and thermography to establish the sympathetic effects of acupuncture. Currently we are exploring and documenting a vast variety of painful conditions to visualize thermographic patterns.

I (ML) had the pleasure many years ago to visit Dr. Sumio Uematsu's laboratory at Johns Hopkins Hospital to reflect at its present state of utilization. We had the pleasure of establishing the first Special Interest Group in Computerized Thermography in the American Academy of Physical Medicine & Rehabilitation, and Dr. Jeffrey Cohen serves as the founding Chairman. Also Dr. Cohen has chaired national and international conferences on thermography.

Our laboratory, the Kathryn Walter Stein Chronic Pain Laboratory, was generously funded by the late Kathryn Walter Stein. Our Attending staff give their time to monthly meetings. Truly a labor of love.

In addition to studying chronic pain and other rehabilitation conditions, thermography offers an expanding horizon of clinical applications into the fields of psychiatry, dermatology, neurology, arthritis, sports medicine, and legal medicine.

Clearly, thermography will be a powerful medical tool and its contribution to health care is at its infancy stage of utilization.

Gratitude:

- Nam June Paik — the late video artist, dear friend, co-panelist, exhibitor and inspiring and delightful mind
- Shigeko Kabota Paik — artist, supporter of rehabilitation medicine
- Robert Perless — for insight and vision, sculptor, patient and friend
- James Goodman — generous support and provision of gallery exhibit space
- Colby Collier and Robert Sablowsky — for securing Kathryn Walter Stein grant to establish and support the thermography laboratory
- Kathryn Walter Stein Family — generous support of our laboratory and research
- Dr. Masayoshi Itoh — mentor, colleague, and friend, who co-directed our initial thermography exploration at Goldwater Memorial Hospital
- Contributing Authors — who took time from their busy schedules to write
- Impress Publications, Publisher — gracious, delightful to work with in editing the book
- Our families — who share the trials, tribulations, and enormous time necessary to birth a textbook

The confluence of technology and physiology is exciting. We hope that this book will spur studies in the field of thermography, with the ultimate goal of enhancing the diagnostic and prognostic qualities of the clinician, thus improving health care.

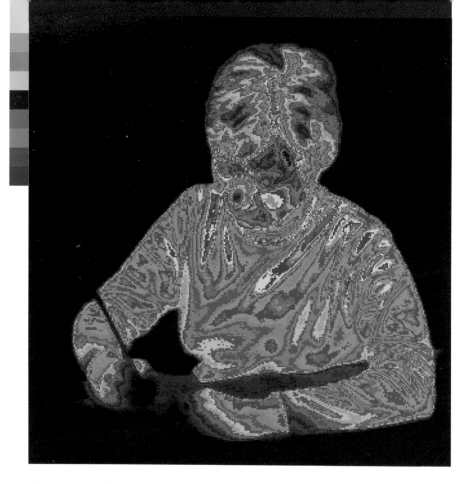

Thermographic image of the author, Barbara Novack

Searching for Pain

by Barbara Novack

Infrared images
red hot to cool blue
swirled in life colors' terrain
energy visible.

Pain alters the image.
Pain alters the being.
Temperatures change:
Paths of pain
made visible.

In the scientific search
pictures are taken
of heat and cold
showing places full glowing,
pulsing with red-orange-yellow warmth and
places where the life force circulates less,
cooling to blues.

There is Design
and there is design.
It is all in
how you look at it.
In the search for pain
the scientist plans and studies,
analyzes,
makes a model and a machine,
takes heat pictures of patients.
And he sees
in his search
mystery
and beauty:
we are art by Design,
but rarely
is the mysterious design
visible.

And even more rarely
does the scientist have
an artist's eye.

Barbara Novack's award-winning poetry has appeared in diverse literary journals, anthologies and three books of collected poems. She has also written six novels and a collection of historical biographies. Ms. Novack, who teaches writing at Molloy College and conducts writers' workshops in the New York metropolitan area, is a member of The Authors Guild, is listed in the Directory of American Poets and Fiction Writers, and is co-founder of the Mendicant Order of Poets (www.mendicantorderofpoets.org), a forum for contemporary poetry. Visit her at www.barbaranovack.com.

"Searching for Pain" was first read on December 9, 2003, at The Union League Club in New York City during a presentation on *The Power, Beauty, and Vision of Thermography* hosted by The Kathryn Walter Stein Chronic Pain Laboratory of The New York University School of Medicine.

Contents

Contributors

Adeel Ahmad, MD, Spine Medicine Fellow, Cleveland Clinic Foundation, Center for Spine Health, Cleveland, Ohio

Izumi Nomura Cabrera, BM, MA, MD, Kaiser Permanente, Oakland, California

Jeffrey M. Cohen, MD, Clinical Associate Professor, Department of Rehabilitation Medicine, New York University School of Medicine, New York; Attending Physician, Rusk Institute of Rehabilitation Medicine, New York University School of Medicine, New York; Medical Director, Kathryn Walter Stein Chronic Pain Laboratory, Rusk Institute of Rehabilitation Medicine, New York University School of Medicine, New York, New York

Laura J. Downing, BS, Research Technician, Kathryn Walter Stein Chronic Pain Laboratory, Rusk Institute of Rehabilitation Medicine, New York University School of Medicine, New York, New York

Michael I. Jacobs, MD, Clinical Associate Professor of Dermatology, Weill Medical College of Cornell University, New York, New York

Mathew H.M. Lee, MD, MPH, Howard A. Rusk Professor of Rehabilitation Medicine and Chairman, Department of Rehabilitation Medicine, New York University School of Medicine, New York; Director, Kathryn Walter Stein Chronic Pain Laboratory, Rusk Institute of Rehabilitation Medicine, New York University School of Medicine, New York, New York

Marc Liebeskind, MD, JD, Clinical Assistant Professor, Department of Radiology, New York University School of Medicine, New York, New York

Sandra H. Moon, MPH

Terri Ann Norden, DDS, MD, Clinical Instructor, Department of Rehabilitation Medicine, New York University School of Medicine, New York; Director of Rehabilitation, North Shore Sports Institute, Great Neck, New York

Bryan J. O'Young, MD, Clinical Associate Professor, Department of Rehabilitation Medicine, New York University School of Medicine, New York; Attending Physician, Rusk Institute of Rehabilitation Medicine, New York University School of Medicine, New York; Visiting Professor, Department of Rehabilitation Medicine, Peking University School of Medicine, Peking First Hospital, Beijing, China; Visiting Professor, Department of Rehabilitation Medicine, Capital University School of Medicine, Xuan Wu Hospital, Beijing, China; Adjunct Professor, Department of Rehabilitation Medicine, University of the Philippines, Philippine General Hospital, Manila, Philippines

Ram C. Purohit, DVM, PhD, DACT, Professor Emeritus, Department of Clinical Science, College of Veterinary Medicine, Auburn University, Auburn, Alabama; Professor, Department of Clinical Sciences and Biomedical Science, School of Veterinary Medicine, Tuskegee University, Tuskegee, Alabama. (All correspondence to: 761 Kentwood Drive, Auburn AL 36830, rpurohit1336@charter.net)

Edwin F. Richter III, MD, Clinical Associate Professor, Department of Rehabilitation Medicine, New York University School of Medicine, New York; Director of Physical Medicine & Rehabilitation, The Stamford Hospital, Stamford, Connecticut

Jay A. Rosenblum, MD, Clinical Assistant Professor, Department of Neurology, New York University School of Medicine, New York, New York

Robert G. Schwartz, MD, Piedmont Physical Medicine & Rehabilitation, PA, Greenville, South Carolina

Sam S.H. Wu, MD, MA, MPH, MBA, Margaret Milbank Bogert Chair of Physical Medicine and Rehabilitation and Medical Director of ICD-International Center for the Disabled, New York; Assistant Clinical Professor, Department of Rehabilitation Medicine, Columbia University College of Physicians and Surgeons, New York; Visiting Assistant Professor, Albert Einstein College of Medicine of Yeshiva University, Bronx, New York; Adjunct Assistant Professor, New York University School of Medicine, New York, New York

The Clinical Use of Temperature Measurement in Medical Practice: A Historical Perspective

SAM S.H. WU, MD, MA, MPH, MBA

An essential condition for human survival is the maintenance of body heat. As a homeotherm, human beings have the ability to maintain constant core body temperature by regulating heat dissipation through controlling blood supply to their skin.[1]

Life is universally associated with a warm body and death with a cold one. Likewise, a moderate body temperature is associated with health and high temperature with illness.[1] This relationship of high body heat and disease has been well known since the dawn of medicine.[5] The study of body temperature, therefore, has its origins in prehistory.[1]

In 1930, the University of Chicago published one of the most ancient medical texts — the seventeenth century B.C. Egyptian papyrus that was discovered at Luxor by Edwin Smith. It described the handling of a wound that was inflamed and suppurating. The instruments of that time were the scanning capacity of the practitioner's fingers and the calculating capacity of the practitioner's brain. "They interpreted and reported that the temperature did indeed rise or elevate over the days, localized in a specific wound or tumor or generalized over the entire body." It is surprising that despite their civilization's preoccupation with quantification in areas such as civil engineering, these ancient healers made no attempts to quantify temperature.[2]

It was not until the time of the Greek pre-Hippocratic medicine (600–400 B.C.) that assessment of body temperature became an integral part of a healer's practice. However, it was Hippocrates (460-379 B.C.) and his followers in *The Book of Prognostics* who first emphasized the diagnostic importance of this temperature assessment. Hippocrates is considered to be the father of modern medicine. He, too, assessed the temperature of his patients using the scanning ability of his hand.[1] In addition, Hippocrates was said to have diagnosed a tumor by applying wet mud on the patient's body and localizing it underneath an area of rapidly drying mud while other areas of mud remained wet.[5]

A half millennium later, the works of Galen (A.D. 130–210) elucidated that body heat was generated from the biocombustion of food. He also recognized the existence of the sensory motor feedback mechanism that we now know is the basis of thermoregulation.[1]

The next major technological leap in the study of body temperature came after the stagnation of the Dark Ages. The glassblowing skills of the Italian artists of the seventeenth century enabled Galileo Galilei to develop the thermoscope in 1595. Open on one end with a bulb on the other, this simple glass tube allowed Galileo to measure changes in temperature regardless of the reference point used.[1,2,4,5,6]

In 1611, Santorio Sanctorius, a professor of anatomy at the renowned Padua University and a friend of Galileo, converted the thermoscope to a quantitative thermometer by making 110 equidistant markings on the glass tube referencing it to melting ice. Using this new instrument, Sanctorius recorded variations in human core body temperature as it related to health and disease.[1,2,4]

Shortly after the introduction of the Fahrenheit and Celsius scales, George Martine in 1740 published his data on the normal temperature of humans.[2,4] It took almost another century and a half before a clinical thermometer was developed by Carl Wunderlich.[2] In 1898, Wunderlich published *On the Temperature in Disease*, in which he compared the body temperatures of healthy and sick individuals.[2,5] In this treatise, he argued for the routine use of the thermometer to measure temperature as part of caring for the infirmed.[5] Despite this seminal clinical work, Wunderlich was ostracized by his contemporaries for this so-called unethical practice.[4] Thus, temperature measurement as a vital sign in clinical practice languished for decades. In the 1930s, Knaus renewed interest in this area by establishing the modern method for routine measurement of body temperature in caring for patient. However, it took another two decades before this method achieved general clinical acceptance in 1952.[1]

In 1800, Sir William Herschel, a British astronomer, discovered the heating power of the infrared rays of the sun.[2,3,4,6] His son, Sir John Herschel, in 1840 rendered these invisible infrared rays visible on specially prepared paper using the evaporograph technique.[2,3,4,5,6] He called the resultant picture a thermogram, which is a term still commonly used today.[5,6]

In 1934, Hardy demonstrated that the skin of human beings emitted infrared radiation as if it were a black body radiator.[4,6] This discovery led to subsequent studies confirming that the skin is a highly efficient radiator of energy from the human body.[4]

Early development of infrared technology was carried out by the military during World War II and became available to industry and medicine in the late 1950s.[2,5] In 1957, R. H. Lawson at the Royal Victoria Hospital in Montreal established the field of medical thermography.[6] This event signaled the beginning of the modern age of infrared imaging.

One of these early instruments was the Pyroscan, which was first used clinically in 1959 to image the heat over arthritic joints. Each picture took several minutes to acquire and was nearly impossible to quantify.[5] Also in 1959, Astheimer and Wormser described an infrared camera that displayed energy vibrations as a gradient of gray densities on photographic film.[6]

During the 1960s and 1970s, systems using liquid crystals as well as remote infrared radiation detectors were developed and refined for clinical use.[1,5] The strengths of the liquid crystal thermographic system were its portability and relative low cost. However, this system was hampered by its poor reproducibility of results due to the inherent need for the liquid crystals to be placed in contact with the skin in order to render an image. The skin surface temperature cannot be reliably measured by any device that makes contact with the skin because the skin's low heat capacity causes its temperature to change on contact with a cooler or warmer object. Furthermore, the liquid crystal thermographic system has poor sensitivity and poor display resolution.[1] The remote infrared radiation detector has higher reproducibility as its images are rendered without any need for contacting the skin. The introduction of oscilloscope and electronic isotherms improved its display resolution.[1,5] However, remote infrared radiation detectors had a significantly higher cost and were not portable until recently.[1]

Thermography — also referred to as Computerized Infrared Imaging or CII — was born in the 1970s when mini-computers were connected to the remote infrared radiation detectors to process the images. In the past three decades, the processing power, portability, and display resolution improved dramatically as the detectors became ever more sensitive and the electronic components became ever more miniaturized. Recently, thermography systems utilizing focal plane arrays have further improved imaging quality. These systems are also capable of capturing images with a high speed and high degree of temperature accuracy.[5] In addition, detector components now do not require liquid nitrogen for cooling the sensors. These advances increased portability and lowered the operating cost of the system. There are thermography systems on the market now that can easily be carried and operated with one hand.

In our nearly four-thousand-year journey from the seventeenth century B.C. to the present, we have seen remarkable advancements in the clinical use of temperature measurement for patient care. The simple touch of the healer's hand is now replaced by the detailed pictures captured by a thermography system. As technology continues to advance at a record pace in areas such as computer science and nanotechnology, the future of thermography in medical practice is indeed bright.

References

1. Anbar M, Gratt BM, and Hong D. Thermology and facial tele-thermography. Part I: History and technical review. *Dentomaxillofac Radiol*, 1998 Mar;27(2):61-7.
2. Bar-Sela A. "The History of temperature recording from antiquity to the present," in *Medical Thermology*, M. Abernathy and S. Uematsu, Eds., American Academy of Thermology, Georgetown University Medical Center, Washington, DC, 1986;1-5.
3. Maxwell-Cade C and Eng C. Principles and practice of clinical thermography. *Radiography*, 1968 Feb;34(698):23-34.
4. Ring EFJ. Quantitative thermal imaging. *Clin Phys Physiol Meas*, 1990;11 Suppl A:87-95.
5. Ring EFJ. The historical development of thermal imaging in medicine. *Rheumatology*, 2004 Jun;43(6):800-2.
6. Winsor T and Winsor D. The noninvasive laboratory: History and future of thermography. *Int Angiol*, 1985 Jan-Mar;4(1):41-50.

History of the Kathryn Walter Stein Chronic Pain Laboratory

JEFFREY M. COHEN, MD
LAURA DOWNING, BS
MATHEW H.M. LEE, MD, MPH

The diagnosis of chronic pain and its myriad of treatment approaches have been major medical, economical, and social issues for decades. Billions of dollars are spent annually on the alleviation of chronic pain problems. As pain has emerged as a devastating public health problem, the need for a laboratory with a focus on chronic pain became strikingly evident.

In response to this need, the *Chronic Pain Laboratory* was established in 1975 by Dr. Mathew Lee in the Department of Rehabilitation Medicine at Goldwater Memorial Hospital in New York. Now located at the Rusk Institute of Rehabilitation Medicine, New York University Medical Center, it was dedicated as the *Kathryn Walter Stein Chronic Pain Laboratory* in 1999.

The primary mission of the *Kathryn Walter Stein Chronic Pain Laboratory* is to address the chronic pain problems faced by patients undergoing rehabilitation therapy. The *Laboratory* uses thermography, also known as computerized infrared imaging (CII), as a diagnostic tool. It provides thermography services to referring physicians. Through the study of the epidemiology of pain and the asymmetric temperature patterns reflected with

The Chronic Pain Laboratory at the Rusk Institute of Rehabilitation Medicine was named for Kathryn Walter Stein following a bequest from the estate of Miss Stein. At the plaque unveiling are, from left to right: John Harney, Mathew H.M. Lee MD, Robert Sablowsky, and David Hellerstein MD.

various pain complaints, thermography has been identified as a useful, objective, non-invasive tool in pain management. The use of thermography on a continuous dynamic basis offers a powerful diagnostic tool for the study of pain, acupuncture, and soft tissue damage.

Thermography detects and records heat radiating from the skin surface. The sensor in the infrared camera is extremely sensitive to temperature variations and displays areas of radiating skin temperatures as specific color regions on a computer screen. The images generated are not pictures of pain, but rather pictures of vascular changes in the body in response to pain. The anatomical areas of interest are subsequently analyzed and average temperature values are generated. Current analytical procedures compare temperatures between symptomatic areas of the body with corresponding contralateral sides of the body.

Goals and Responsibility

Since its inception, Dr. Mathew Lee has provided the leadership towards achieving the purpose and goals of the laboratory. These goals are:
- to study the epidemiology of pain and patterns
- to study the various clinical aspects of the rehabilitation of chronic pain patients
- to develop clinical models of teaching the diagnosis and management of chronic pain
- to utilize thermography as a diagnostic and therapeutic measuring tool in studying chronic pain patterns
- to observe the clinical effects of acupuncture for pain abatement via thermography.

Based on these goals, the following triad of responsibility was created in order to fully address all aspects of chronic pain:

Triad of Responsibility		
A **Clinical Care Diagnosis and Treatment**	**B** **Education**	**C** **Research**
Examine and review difficult chronic pain patients	Participate in training of residents and students in the health professions for chronic pain. Conduct monthly meetings to discuss cases and current literature	Encourage studies and publications in the field of chronic pain

In addition to clinical pursuits, the *Kathryn Walter Stein Chronic Pain Laboratory* staff is continuously active in thermography research. The computerized infrared imaging technology enables one to assess qualitative and quantitative changes in cutaneous temperature, reflecting underlying sympathetic activity. This allows one to study a variety of sympathetic mediated diseases and autonomic neuropathies. The staff of the *Kathryn Walter Stein Chronic Pain Laboratory* has published and presented numerous papers and posters at national and international scientific meetings on the application of thermography in a rehabilitation setting.

The laboratory has also made significant contributions to the understanding of acupuncture analgesia, demonstrating a relationship between acupuncture and the autonomic nervous system using thermography. Over the past 30 years, Dr. Mathew Lee has advanced the standards of acupuncture practice and has guided educational and research efforts in acupuncture.

With the founding of the *Chronic Pain Laboratory*, Dr. Lee applied his knowledge and study of acupuncture to the technology of thermography. The *Laboratory* has been cited for its seminal studies elucidating the sympatholytic effect of acupuncture visualized through thermography.

Some of the laboratory's other research milestones include its work involving:
- dental analgesia/naloxone dental analgesic experiments
- tooth pulp stimulation studies with silicone rubber mold
- laser stimulation evoked potential studies at the Technion University/Israel.

Current Research

As the thermography technology continues to improve in sensitivity, analysis capabilities, and portability, the laboratory is continuing to expand its research efforts. Currently, research areas include:
- observing temperature differences in wounds such as open sores, burns and decubitus ulcers
- evaluating circulation patterns in patients with diabetes
- assessing circulation changes in relation to pain complaints in stroke patients
- observing circulation changes in patients with Complex Regional Pain Syndrome/Reflex Sympathetic Dystrophy
- researching the changes in temperature of the hands of professional musicians before and after playing their instruments
- recording the physiologic effects of acupuncture.

These areas of research are providing the foundation for the establishment of thermography in the rehabilitation setting.

For the future, the *Kathryn Walter Stein Chronic Pain Laboratory* has set high goals in the area of chronic pain. Some of the future plans include:

- applying the thermographic technology on a larger scale throughout the hospital with the use of a portable infrared camera. With this capability, the lab can provide regular assessments to patients during their hospital stay, evaluating decubitus ulcers (bed sores) and wound healing.
- providing outpatient thermographic evaluations, both within the hospital and in various other locations. Because infrared technology is a non-invasive test, serial evaluations may be utilized as a prognostic tool in pain and/or injury prevention.
- continuing the utilization of thermography in documenting the physiological effects of acupuncture and other non-traditional forms of medicine.

Physiology of the Skin

MICHAEL I. JACOBS, MD

The science of Thermography is better understood if one has a basic knowledge of the structure and function of the skin, its vasculature, and its nervous system. It is quite interesting that the skin provides not only clues to diagnose systemic diseases, but also is a window to let us monitor the health of our blood vessels and nerves. It is for this reason that thermography, a scientific technique that expands the diagnostic ability of our eyes, is such an exciting field.

The skin is composed of three components. The top layer is the epidermis, below this the dermis, and finally the fatty or adipose tissue, which is anchored to the muscle fascia. The thickness of the skin varies with different locations on the body, being usually thickest on the back, and thinnest on the eyelids.

Epidermis

The epidermis is the component of skin we normally see with the eye. It is the thinnest component, and when patients have porcelain white almost translucent skin, one can easily see the blood vessels and pink hue underneath it, which are components of the dermis. The epidermis provides a barrier to the outside world, and as such it helps keep constant the fluid and electrolyte balance of the body. This barrier function helps to prevent penetration of microorganisms, and, aided by the immune cells in the epidermis, to fight infections.

The epidermis is separated from the dermis by a membrane, termed the "dermal-epidermal junction." This membrane is composed of fibers that anchor the epidermis to the dermis, and also a matrix containing many important biochemical components including laminin, fibronectin, and collagen. The dermal-epidermal junction regulates the permeability between the epidermis and the dermis to nutrients, immune molecules, and even cells. The entire nutrition of the epidermis comes from the dermis, as the epidermis contains no blood vessels.

Dermis

The dermis is the component of the skin that contains the connective tissue, blood vessels, and nerves. Along with the epidermis, the dermis protects the body from physical injury, helps retain water and nutrients, and through the nervous and vascular systems, helps regulate the temperature of the body. For scientific purposes, it is divided into a superficial papillary layer and a deeper reticular layer. While the epidermis is composed mostly of cells termed keratinocytes, the dermis mainly consists of connective tissue and amorphous matrix between the cells. Sitting within the connective tissue and amorphous matrix are the glands, the nervous system and the cells. The matrix of the dermis consists of collagen, elastin, glycoproteins, proteoglycans, and glycosaminoglycans.

The majority of the dermis is made up of collagen. Collagen and elastin provide the skin with its strength and resilience. Most of the collagen molecules are Type 1 and Type 3, although Types 4–7 play important roles. The elastic network of connective tissue extends throughout the dermis, within the walls of the blood vessels of the skin, and around the hair follicles. The network of elastic fibers returns the skin to its normal shape after stretching or compression. Many cells of the blood circulate through the dermis including monocytes, macrophages, dermodendrocytes, and mast cells. These cells are involved in the immune function of the skin, allergic reactions, and body defenses against infection and tumor surveillance.

The vasculature of the dermis is divided into two regions, the papillary dermis and the reticular dermis. The papillary dermis resides immediately below the dermoepidermal junction membrane, and the reticular dermis is located below the papillary dermis and extends down to the subcutaneous adipose tissue. The vascular networks are divided into the superficial and deep plexus of arterioles and venules of the skin. These are connected by ascending and descending blood vessels. The larger arteries and veins are contained in the fatty tissue layer of the subcutaneous skin. All of the vessels communicate. The superficial plexus is located just beneath the papillary dermis. From the superficial plexus, capillaries loop vertically into each dermal papilla. The upward arterial component wanders through the dermal papillae to the most superficial aspect, where it changes into a postcapillary venule and returns back to the sub-papillary level. The deep plexus of arterioles and venules is located in the bottom part of the dermis. There is an abundant interconnection between the superficial and deep vascular plexuses. It is difficult for blood flow to be blocked in the skin, as there are so many communications between these vessels. The walls of the blood vessels themselves are thickest in the deeper subcutaneous fatty tissue, and thinnest in the superficial dermis and papillary capillaries. The walls of the capillaries and postcapillary venules are permeable to nutrients, water, oxygen, and hormones. Metabolic waste products are collected by the postcapillary venules and transported back to the body. During inflammatory responses, vasoactive chemicals are released causing increased permeability of the postcapillary venules, resulting in fluid extravasation. This can lead to redness of the skin and skin swelling. The prototype for this would clinically be termed a "hive" or urticaria.

The skin has a structural mechanism to allow blood to be shunted away from the papillary dermis and remain deeper in the skin. These shunts are termed glomus bodies, which direct blood from arterioles to venules, circumventing the capillary bed. When the capillaries are bypassed, the speed and volume of blood flow increases. The distal extremities such as the fingers, toes, and also the ears and nose contain the highest number of glomus bodies. The glomus bodies contain

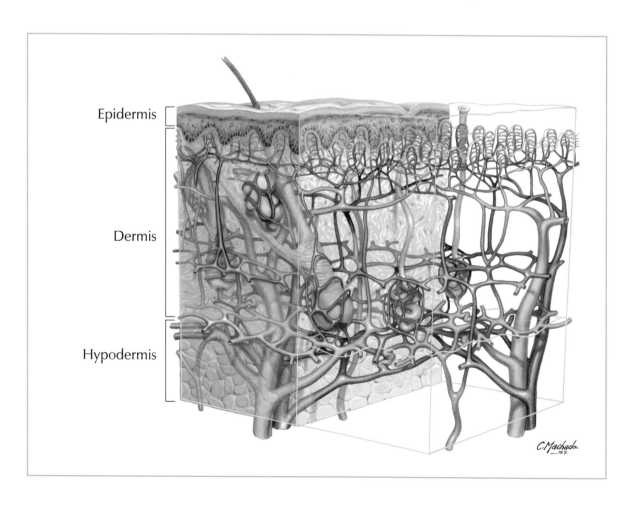

Epidermis

Dermis

Hypodermis

modified smooth muscle cells which are innervated by nerves of the sympathetic nervous system.

The control of the flow of blood through the skin helps to regulate body temperature and blood pressure. The requirements of the skin for nutrients are quite small compared to the total volume of blood which flows through its vessels. Under normal conditions, approximately 5–10% of the cardiac output goes through the skin. This can increase up to 7 times or decrease to almost 0 in conditions of extreme external temperatures. The sympathetic nervous system maintains the tone of the blood vessels, which helps us maintain our blood pressure and blood flow during upright posture, exercise, and temperature change.

The control of blood pressure takes precedence over control of body temperature. Arterioles are vessels of resistance, and venules are vessels of capacitance. The primary control is through the adrenergic sympathetic nervous system. Cardiopulmonary receptors monitor pressure, and sensory receptors in the skin monitor temperature regulation of the extremities. In addition, hypothalamic receptors help monitor core body temperature.

The skin is extremely sensitive to temperature change. The nervous system in the skin can distinguish between hot and cold with good accuracy. Variations in temperature of a fraction of a degree can be appreciated. Core body temperature is maintained between 36.2 and 38 °C. Core temperature is a result of the heat generated within the body and its loss to the external environment. The metabolic chemical reactions required by the body occur at the optimum core body temperature. When the temperature of the external environment increases, the body tries to lose heat through the skin. The cutaneous vasculature transfers heat produced by the body core to the external environment. This is controlled by sensory receptors in the hypothalamus and spinal cord. When the core body temperature is elevated due to a rise in temperature in the external environment, more blood is shunted to the surface of the skin and blood flow increases. When the external environment becomes cold, the skin helps to preserve core body temperature through decreased blood flow to the surface by reflex vasoconstriction.

It is important that while temperature appreciation by cutaneous thermoreceptors results in localized adaptation to the external environment, it is the central core body temperature as set by the hypothalamus that controls the flow of blood in the cutaneous vessels of the entire body. Dissipation of heat through convection and conduction account for a minor fraction of total cutaneous heat loss. The majority of heat loss is

> The skin is extremely sensitive to temperature change.

through infrared radiation, followed by evaporation through water loss and sweating. Localized changes in skin temperature can be induced by acupuncture. Thermography is the perfect tool to monitor this effect.

Hypodermis

In the bottom-most layer of skin, the hypodermis, the adipose tissue is the main component. Fat fills the gaps between connective tissue anchors, located between the bottom of the reticular dermis and the muscle fascia. This adipose tissue contains the germinative areas of the hairs, apocrine and eccrine sweat glands, nerves, blood vessels, and lymphatics. As mentioned, the blood vessels of the subcutaneous fat are the largest ones found in the skin, and have the greatest volume and flow of blood. The fat cells themselves also provide a supply of energy, and have the physical ability to cushion the skin against underlying structures and the muscle.

The Skin and Nervous System

The complexity of the cutaneous nervous system serves to inform, protect, and regulate the body's response to the outside world. Recent evidence even points to a connection between the nervous system and the body's immune system. The nerves of the skin are mainly divided into two categories, sensory and autonomic. The sensory nerves, through interpretation by the brain, are responsible for transmitting the sensations of temperature, pain, itch, light touch, pressure, vibrations and proprioception. The autonomic nervous system controls the tone of the cutaneous blood vessels, and the state and activity of the hairs and glands in the skin.

Just as the larger blood vessels enter the skin through the subcutaneous fatty layer, the same is true for the cutaneous nerves. From their origins in the spinal cord and brain stem, the nerves enter the subcutaneous level and ascend to form a deep and superficial plexus of nerves. The more superficial in the skin, the smaller the nerve bundles until they reach the papillary dermis where the smallest nerves can be appreciated. Cutaneous sensory nerves are oriented to transmit signals along dermatomal pathways back to the central nervous system. Autonomic nerves travel from sympathetic ganglia, mingle with sensory nerves, and terminate in the eccrine glands, apocrine glands, cutaneous blood vessels and muscles associated with erection of hair fibers in the skin.

There are many specialized nervous system receptors in the skin to perform these functions. Neurofibers in the skin are both myelinated and unmyelinated. Myelinated fibers have a higher conduction velocity. Autonomic fibers are unmyelinated and as opposed to capillaries, which are present in the papillary dermis but do not penetrate the dermal-epidermal junction, intraepidermal free nerve endings are present in the epidermis. An interaction between these nerve endings and immune cells of the epidermis termed Langerhans cells has been noted. Small diameter myelinated and unmyelinated fibers control sensations of pain and temperature. Cutaneous blood vessels are innervated by unmyelinated, sympathetic postganglionic nerves. The apocrine and eccrine glands in the skin are supplied by unmyelinated adrenergic and cholinergic nerves. The vascular smooth muscle contains receptors activated by adrenergic fibers.

The cutaneous sensory receptors can be divided into two groups. One group consists of specialized receptors, which are encapsulated. The other is composed of non-encapsulated receptors or free nerve endings. Most of the sensory nerves in the skin are in this latter group. Nerve endings associated with Merkel cells are also in this category. Merkel cells are thought to function as touch receptors and also may have an immunologic role in the skin. The encapsulated skin receptors are mechanoreceptors. These include Pacini corpuscles, which are chiefly situated on weight bearing surfaces in addition to being plentiful in the lips, genitalia and nipples. Meissner corpuscles are located in the tips of the dermal papillae on the hands and feet, and are most concentrated on fingertips. They are thought to function as receptors, which detect the sensation of light touch. Mucocutaneous end organs and genital corpuscles are located on the borders of mucus membranes such as the lips of the face, eyelids, glans penis, prepuce, clitoris, labia minora, and perianal skin. Hair follicles are innervated by extensively branched nerve fibers. Sensations appreciated by the brain as coming from the skin include touch, pressure, vibration, and spacial discrimination, appreciation of temperature, pain, and itch.

In sum, basic knowledge of the skin and its functions will provide insight into the capability and interpretation of the scientific results of thermography. Diseases affecting the vascular system, nerves, and connective tissue will result in temperature changes detected by thermography. Pharmacologic manipulation, inflammation and other agents affecting thermography require the interpretation of the scientist who understands the basic structure and workings of the skin. It is my wish that this chapter provide the spark of interest for expanded appreciation of the plethora of functions of and disease affecting the skin.

Thermography Techniques

Izumi Nomura Cabrera, BM, MA, MD
Jeffrey M. Cohen, MD
Laura Downing, BS

Proper thermography technique is critical to the validity and utility of infrared images in the clinical setting. Factors that must be addressed in order to produce quality images include the equipment, the room in which imaging takes place, and the preparation and positioning of the patient.

Appointment Preparation

Proper thermography technique begins as soon as the patient calls for an imaging appointment. The appointment secretary must obtain information regarding the area(s) of pain, as well as the referring physician's name. All appointments must have a referral from a physician and cannot be scheduled without one. The patient is informed that the report from the imaging test will be sent to the referring physician when completed.

Prior to the appointment, the patient is instructed to refrain from ingesting any food, drink, or medication for 3 hours prior to the exam, unless contraindicated by the referring physician. In addition, no strenuous physical activity, such as running, aerobics, swimming, or physical therapy should be performed approximately 6 hours prior to the exam. These requirements involve careful planning when scheduling a patient who may have a particular itinerary of meals and activities to work around.

The patient must also refrain from using any ointments, lotions, hot pads, ace bandages, or splints for 6 hours prior to the exam, unless contraindicated by the physician. Cosmetics (if facial imaging) and perfumes must also be withheld on the day of the imaging appointment, as they can mask the skin's true temperature.

At the time of scheduling the appointment, the patient is informed that the test requires staying in one position for 15 minutes without moving. If the patient feels this restriction is impossible due to physical limitations and/or pain, the appointment must be rescheduled. In addition, the patient is asked to inform the imager if there is any discomfort at any time during the test. (Due to these vital instructions at the time of the testing, it is imperative that the patient be able to comprehend the verbal information. If there is a language barrier present, an interpreter must be present during the exam.)

During the Appointment

At the time of the appointment, the patient fills out a questionnaire regarding the history of the present illness/injury. No physical examination is conducted. The history is reviewed by the technician, and a graphical description of the patient's pain is drawn using a Graphical Description of Pain form. Any additional painful areas are noted. The test procedure is fully explained, with the reassurance that the test is noninvasive.

Equipment

There are many types of thermography systems available on the market today. They range from the larger, more traditional models connected to a desktop computer to the newer, portable handheld models.

The main system used by the Kathryn Walter Stein Chronic Pain Laboratory is a Thermal Image Processor System (Computerized Thermal Imaging Inc., Ogden, Utah). This system has three components: a high resolution thermal imager with closed cycle cooler, a proprietary computer interface card, and a personal computer.

The thermal imager has a twelve-sided polygon horizontal scanning mirror. As this mirror rotates, it picks up information from the field of view. This information is then gathered on an elliptical vertical scanning mirror and reflected onto a single element 60 micron Mercury Cadmium Telluride detector.

- The standard CTI camera is sensitive to infrared radiation in the 8 to 12 micron range.
- The camera has a temperature range of 50 °C, achieving a resolution of 0.0125 °C.

The thermal data is converted into a digital signal, one pixel at a time, and transferred to the computer interface card. This data is then assembled into a frame of data and passed on to the computer.

Taking Thermography Images

The sequence of imaging is based on the following:
1. Location of pain
2. Number of areas to be imaged
3. Comfort of the patient.

The patient is asked to remove all restrictive clothing, particularly in the areas to be imaged. Clothing in areas not pertinent to the exam must be loose, allowing for blood flow to remain unobstructed. Following appropriate patient disrobing, the areas to be imaged are examined to make sure that they are free of any bandages, ointments, jewelry, etc. and are completely exposed.

Figure 1 (above): *A photo of the Thermal Image Processor System. On the left is the high resolution thermal imager and on the right the personal computer. The proprietary computer interface card is not visible.*

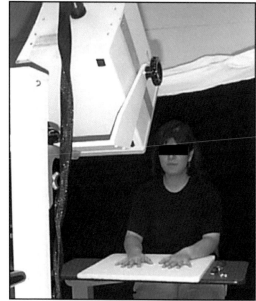

Figure 2 (right): *An alternate view of the system, showing patient positioning for dorsal hand and wrist imaging.*

The number of areas to be imaged is determined based on the location of pain. Positioning is performed to obtain unobstructed views of the required areas with a minimal amount of patient discomfort. For this reason, a stretcher can be particularly helpful as the patient's body can be assessed in the sitting, trendelenburg, and reverse trendelenburg positions.

Pillows and other types of supports can be used to help minimize patient discomfort. Often the technician will realize that numerous positions will be necessary, requiring additional time from the patient. Therefore, when the sequence of imaging has been decided, verification of overall time needed to complete the session is conferred to the patient.

If completion of imaging is not possible due to time constraints, the patient is asked to reschedule as a time constant is preferable.

In order to minimize changes in cutaneous temperature due to environmental factors and/or muscle activity, the following guidelines are followed:

1. The patient equilibrates to the room temperature without movement for 15 minutes prior to taking the desired thermography image. To do so, the patient is positioned comfortably, with the area to be imaged exposed to room air, and is asked to remain as still as possible for 15 minutes. Large muscle movements and skin manipulations, such as scratching and rubbing the skin are to be avoided.

2. The room is kept temperature controlled at 18-22 °C and is free of any drafts. All windows in the room are covered. These measures minimize environmental factors that may influence the temperature of the skin.

3. The imaging screen is positioned away from the patient in order to prevent any biofeedback effect.

4. The equilibration process is repeated each time the patient is repositioned to expose a different body part.

The images taken must provide the optimal view of the area of interest, ideally directly in front of the camera lens. Utilization of anatomical landmarks to assist in this analysis is helpful. Images are taken of the particular body area of interest and its exact homologous site. The size of the areas imaged depends upon the patient's pain complaints.

Table 1 is an example of a checklist sheet used to determine images to be taken based on pain complaints.

Table 1. *Checklist of images to be taken*

Patient Sitting

Head	3 views	☐ Front
		☐ Left
		☐ Right
Neck	4 views	☐ Front
		☐ Left
		☐ Right
		☐ Back (with hair raised)
Upper Ext.	2 views	☐ Ventral arms 90° abduction
		☐ Dorsal arms 90° abduction
Hand	2 views	☐ Dorsal hand and wrist
		☐ Ventral hand and wrist

Patient Lying Down

Back	2 views	☐ Upper back from head, patient prone
		☐ Lower back and buttocks from feet, patient prone
Lower Ext.	5 Views	☐ Anterior LE from feet, patient supine
		☐ Dorsal feet, patient supine
		☐ Posterior LE from feet, patient prone
		☐ Soles, patient prone
		☐ Lower back and buttocks from feet, patient prone

In addition to the sequence of images based on pain complaints, occasionally an area may appear distinctive on initial thermography impression. If this should occur, an additional image(s) is recommended along with documentation of the area in question.

Analysis of Thermography Images

In order to prevent interpreter bias, the thermography images are evaluated by a blinded reviewer who was not present during the imaging session and has no prior knowledge of the patient's clinical history.

Outcome measures are the presence of asymmetric skin temperature patterns recorded and measured by thermography. Identifying areas of pain with thermography has traditionally been performed through the comparison of one side of the body to its corresponding site on the contralateral side, using the side without pain as the "control." Body areas are then determined as being symmetric or asymmetric to each other with regards to temperature.

Studies have shown temperature symmetry to be well conserved in homologous areas in the absence of pain.[2,3] Uematsu studied 32 normal subjects and 30 patients with peripheral nerve impairment who ranged in age from 12 to 65 years. Uematsu found that in normal persons the skin temperature difference between contralateral sides of the body was only 0.24 +/- 0.073 °C.[2,3] He noted that skin temperature differences between corresponding sites on one side of the body compared to the other were not only extremely small, but also very stable throughout the body. Uematsu concluded that there is minimal temperature variation between corresponding sites on different sides of an individual's body.

In addition to asymmetry analysis, patterns may also be evident. They may be discrete or prominent in nature, and interpretation of these patterns must be performed judiciously. With all impressions, clinical correlations are recommended.

> Studies have shown temperature symmetry to be well conserved in homologous areas in the absence of pain.

Normative Data

Due to the conservation of temperature symmetry in homologous regions of the body, in the absence of pain or in the presence of bilateral pain,[1-2] researchers at the *Kathryn Walter Stein Chronic Pain Laboratory* undertook an alternative approach to the analysis of thermography.[1] A retrospective analysis of thermography imaging data of 110 patients (904 views) was performed. Data was categorized for 28 major body areas, all of which were devoid of pain. Main outcome measures were the cutaneous temperature averages and standard deviations for the 28 major body areas derived from mapped body areas. The objective of this study was to establish a normative database that may be used for comparison, particularly in patients with bilateral pain.

Figures 3 and 4 show anterior and posterior views and the average temperature values and standard deviations of these 28 areas.

Since the data recorded were of areas with no complaints of pain, blood flow within each body area was considered to have not been affected by abnormal sympathetic activity. The data indicated that the average cutaneous temperatures obtained from the 28 body areas varied according to location, with cooler temperatures recorded at the more distal areas. This is expected physiologically due to their relative distance from the body core. This normative database with its locus-specific absolute measurement approach stands in contrast to conventional thermography procedures. It can provide the physician with information not typically observed when analyzing thermography for asymmetry alone.

Figure 3. *Anterior view showing average temperature values*

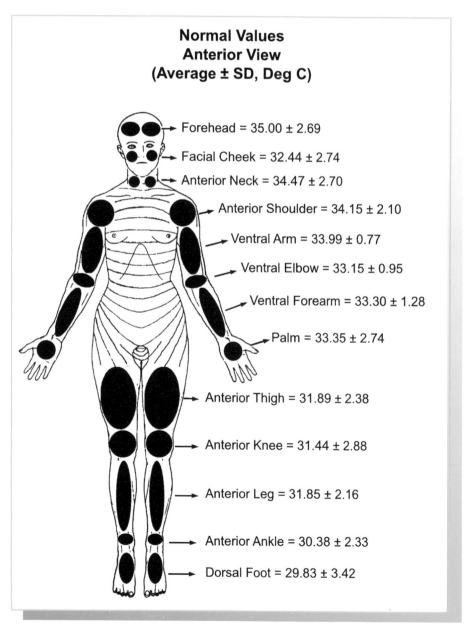

Normal Values
Anterior View
(Average ± SD, Deg C)

Forehead = 35.00 ± 2.69

Facial Cheek = 32.44 ± 2.74

Anterior Neck = 34.47 ± 2.70

Anterior Shoulder = 34.15 ± 2.10

Ventral Arm = 33.99 ± 0.77

Ventral Elbow = 33.15 ± 0.95

Ventral Forearm = 33.30 ± 1.28

Palm = 33.35 ± 2.74

Anterior Thigh = 31.89 ± 2.38

Anterior Knee = 31.44 ± 2.88

Anterior Leg = 31.85 ± 2.16

Anterior Ankle = 30.38 ± 2.33

Dorsal Foot = 29.83 ± 3.42

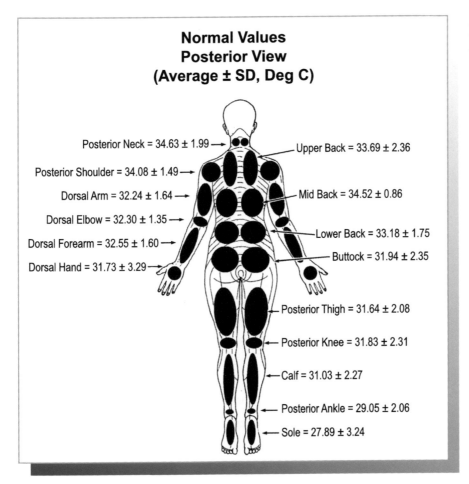

Figure 4. *Posterior view showing average temperature values*

References

1. Cabrera IN, Wu SSH, Haas F, and Lee MHM. Computerized Infrared Imaging: Normative Data on 110 Patients. *Arch Phys Med Rehabil*, 2001;82:1499.

2. Uematsu S, Edwin DH, Jankel WR, et al. Quantification of thermal asymmetry, Part I: Normal values and reproducibility. *J Neurosurg*, 1988;69:552-5.

3. Uematsu S, Edwin DH, Jankel WR, et al. Quantification of thermal asymmetry, Part II: Application in low-back pain and sciatica. *J Neurosurg*, 1988;69:556-61.

Infrared Thermographic Vasomotor Mapping and Differential Diagnosis

ROBERT G. SCHWARTZ, MD

When performed with proper technique and under controlled conditions, thermography (Computerized Infrared Imaging or CII) is the test of choice for mapping of vasomotor instability and asymmetry. The findings provide important clinical insights into those structures that generate aberrant sympathetic responses for pain syndromes such as Reflex Sympathetic Dystrophy (RSD), Complex Regional Pain Syndrome types I and II (CRPS), Thoracic Outlet Syndrome (TOS), Cervical Brachial Syndrome, Fibromyalgia, and Barre-Lieou. In addition, the presence of abnormalities and the distribution of findings can be invaluable in differential diagnosis of these conditions.[1]

The medical community has demonstrated increased awareness of sympathetic pain syndromes over the last decade. New interventions and approaches toward alleviating symptoms in those afflicted have been tried, some with success. Even better results can be achieved through a greater understanding of which structure is initially responsible for generating the condition.

The sympathetic system, which is largely responsible for the control of surface skin temperature, innervates all tissue of mesodermal and ectodermal origin. For non-visceral soft tissues, this includes muscle, ligament, synovium, tendon, fascia, dura, disc, and peripheral nerve fibers. Other less obvious but equally important structures, such as interosseous membrane and neuro-lymphatic sphincters, can be richly innervated by the sympathetic system as well. Essentially, the innervation of the sympathetic system is ubiquitous.[12,19]

Since one of the primary functions of the sympathetic system is to monitor those tissues that it innervates, it is not surprising that when an injury occurs to one of those structures, the system may occasionally function improperly. Why this occurs remains speculative, but the net result is an alteration in transmembrane electric potential of the affected sympathetic nerve fibers. Direct structural injury, vascular ischemia, infection, and coagulopathy are just a few of the mechanisms that might lead to such an alteration.

From a thermographic perspective, what is important is whether the resultant vasomotor response is great enough to create a change in skin temperature of greater then 1 ºC compared to the contralateral side or with respect to the surrounding dermatome, sclerotome, or vasotome. While dermatomes represent the distribution of sensory nerve fibers upon skin, a sclerotome reflects the distribution of skin galvanic impedance influenced by a visceral or non-visceral soft tissue structure. Numerous sclerotomal patterns exist. Examples of clinical conditions

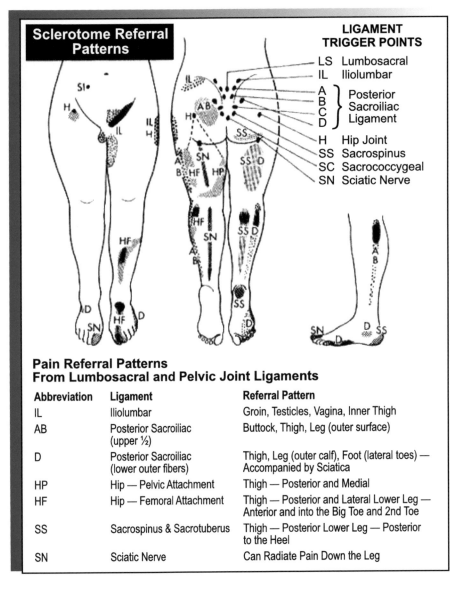

Figure 1: *Lower Extremity Sclerotomal Patterns. Adapted from Hauser, R. Prolo Your Headache And Neck Pain Away, Beulah Land Press, Oak Park, Illinois, 2006.*

Sclerotome Referral Patterns

LIGAMENT TRIGGER POINTS

LS Lumbosacral
IL Iliolumbar
A ⎫
B ⎬ Posterior Sacroiliac
C ⎪ Ligament
D ⎭
H   Hip Joint
SS Sacrospinus
SC Sacrococcygeal
SN Sciatic Nerve

Pain Referral Patterns
From Lumbosacral and Pelvic Joint Ligaments

Abbreviation	Ligament	Referral Pattern
IL	Iliolumbar	Groin, Testicles, Vagina, Inner Thigh
AB	Posterior Sacroiliac (upper ½)	Buttock, Thigh, Leg (outer surface)
D	Posterior Sacroiliac (lower outer fibers)	Thigh, Leg (outer calf), Foot (lateral toes) — Accompanied by Sciatica
HP	Hip — Pelvic Attachment	Thigh — Posterior and Medial
HF	Hip — Femoral Attachment	Thigh — Posterior and Lateral Lower Leg — Anterior and into the Big Toe and 2nd Toe
SS	Sacrospinus & Sacrotuberus	Thigh — Posterior Lower Leg — Posterior to the Heel
SN	Sciatic Nerve	Can Radiate Pain Down the Leg

with identified sclerotomal patterns (often described as referred pain) include facet syndromes, myofascial, ligament, and dural pain syndromes (Figure 1).

It is important to recognize that while sclerotomal patterns frequently mimic pain patterns such as herniated disc, they are not at all pathognomonic for the same. For example, a fibulocalcaneal ligament strain may very well have thermographic change that tracks in an L5 distribution (Figure 2), but that does not mean that the L5 nerve root or disc is the source of those findings. While the nerve root or disc may be the source, all structures that refer within that sclerotome must be considered when deciding which structure is responsible for the abnormality.[5]

Likewise, it is important to understand that treating any structure within the sclerotome may actually correct the abnormality. Sometimes all that is required to restore skin galvanic impedance to normal, and its associated vasomotor instability or asymmetry, is to remove the stimulus that initially generated the sympathetic response. This may mean an injection of a medicine into a torn ligament that stops inflammation or repairs the tear, or of a neurolytic agent that alleviates a persistent non-physiologic contraction of muscle. Naturally, other examples exist, such as hyaluronidase injection into a knee, and oral or topical medications that restore blood flow and modulate sympathetic tone.

Not withstanding the above, there is also every reason to believe that treating cephalad to the most proximal portion involved will be more effective then treating caudal to it. The medical literature is replete with references demonstrating the benefit of spinal blocks for sympathetic pain syndromes. It is not, however, as clear why some blocks are more successful then others.

If a lumbar thermographic study demonstrates vasomotor asymmetry that tracks in an L34 dermatome or sclerotome, it would be reasonable to speculate that a sympathetic block at L3 would be more effective

Figure 2: *Left Lower Extremity tracking cold in a L5 distribution (in this case secondary to a fibulocalcaneal ligament strain).*

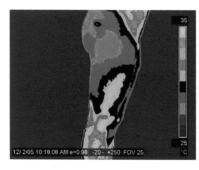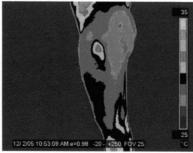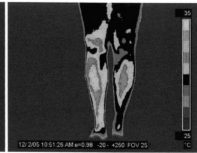

then an epidural block at L5. In addition, if the original injury was in the ankle at the medial collateral ligament of the knee, then a concurrent injection with intensive physical therapy at that locale may prove to be even more rewarding then either intervention alone. By obtaining thermographic imaging, powerful answers as to the extent and distribution of involvement can be obtained.

> While dermatomes represent the distribution of sensory nerve fibers upon skin, a sclerotome reflects the distribution of skin galvanic impedance influenced by a visceral or non-visceral soft tissue structure.

The thermographically generated vasomotor map also provides invaluable information for therapeutic decision-making when treatment previously based upon it fails. For example, if a lumbar block does not produce pain relief in an L5 vasomotor-mapped patient, the patient may still show dramatic response to a peroneal nerve block (another L5 innervated structure). A combination of expertise in the basic anatomy of those structures that can exert influence in the distribution of the vasomotor abnormality found, and the ability to objectify where the vasomotor asymmetry actually occurs, allows for a more rational approach to intervention that is otherwise not available.

Vasotomes represent another pattern of abnormality that the examiner must understand. They should not be confused with vasomotor instability of sympathetic origin. Vasotomes are not dependant upon sympathetic control of skin galvanic impedance, cutaneous vasculature or sweat glands, but rather represent peripheral vascular supply zones.[29]

Likewise, a local inflammatory condition, such as a hot joint in rheumatoid arthritis or erythema associated with a rash, should not be confused with local vasomotor dysfunction under sympathetic influence. By completing a full thermographic study (bilateral extremities from multiple views and corresponding spinal segments), it is not at all difficult to differentiate local inflammatory, venous, or peripheral artery abnormalities from sclerotomal or dermatomal patterns.

A normal study is also clinically helpful. It is not uncommon for a patient to be given a diagnosis of CRPS/RSD and yet be non-responsive to sympathetic block. Prolonged, hopeless medical management or invasive procedures such as spinal cord stimulators can result. A normal study helps rule out the original diagnosis, or at least suggests that a sympathetically independent pain syndrome may exist.

A localized thermographic pattern inconsistent with other recognized patterns can provide useful information as well. For example, when warming is present in the dorsolateral aspect of the foot alone

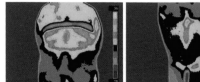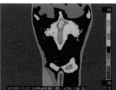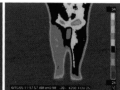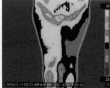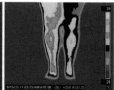

Figure 3: *Note the warm left dorsal foot consistent with an Angry Back Firing C Syndrome. There was no evidence of a heat asymmetry pattern in any other view.*

(Figure 3), the examiner should look for a missed fibulotalo ligament strain that, when treated, may be miraculously responsive. Since sympathetic variants such as the Angry Backfiring C (ABC) syndrome (where a backfiring of the C fiber results in excess Substance P accumulation) may also create a similar picture, differential diagnostic skills must still be employed.[26]

In the case of ABC syndrome, sympathetic block may not only fail, but can create a paradoxical worsening of symptoms, as the painful part is already vasodilated.[20] In this instance a pharmacologic approach that is intended to deplete Substance P or target receptors responsible for vasoconstriction, or employment of electric sympathetic block (where different aspects of the voltage gate can be targeted) may prove more effective then a chemical sympathetic block.[25] Whenever a paradoxical response to sympathetic block occurs, this should be kept in mind.

In addition to objectifying the presence of a paradoxical effect, Infrared Thermographic monitoring during blockade can be quite helpful in assessing if intended ipsilateral vasodilatation was accomplished. Even when a Horner's is observed with a stellate block, as many as 40% of patients do not get limb vasodilatation.[18] Their lack of clinical responsiveness to the block may lead to a false impression that CRPS/RSD does not exist.

The Triple C syndrome, consisting of cold hypesthesia, cold allodynia and cold skin, is another localized sympathetic variant.[20,27] As expected, with this syndrome Infrared Thermographic imaging reveals a localized cold asymmetry pattern. The more distal the occurrence of this syndrome, the less responsive the patient is to a spinal block. With Triple C syndrome, combination interventions, including localized therapy and pharmacologic agents, should be more aggressively used.

Diffuse vasomotor instability involving an entire limb or limb segment (Figure 4), and not confined to a particular dermatome or sclerotome, is a hallmark finding of a true RSD syndrome.[8,32] Dural, neuro-immuno-infectious interactions and multiple generators should be aggressively investigated.[28] While any case of sympathetic pain with

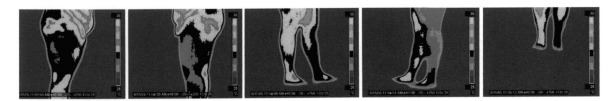

Figure 4: *Heat Emission asymmetry pattern with diffuse left lower extremity involvement consistent with Reflex Sympathetic Dystrophy.*

vasomotor instability can spread, when diffuse vasomotor asymmetry exists, symptomatic intervention with an eye towards prevention of spread, limb trophic changes, edema, contracture or Sudeks atrophy should be emphatically employed.[4,6,14,23]

Stopping progression is one of the most effective treatments a physician has to offer in the treatment of CRPS/RSD. Early diagnosis, due to high sensitivity, is one of the great advantages that thermography offers over triple phase bone scan or diagnostic block in the management of sympathetic pain syndromes.[10,13,22] Findings of diffuse vasomotor asymmetry should alert the physician to intercede promptly to interrupt the progression of CRPS/RSD toward stages two and three.

The physician must keep in mind that thermography is no different than any other objective study. Ultimately, it is always best to treat patients based upon both clinical and diagnostic impressions, not test results alone. This approach will help avoid the potential pitfall case wherein a localized, or clearly defined, asymmetry pattern unexpectedly shows rapid progression to escalating stages.

The International Association for the Study of Pain (IASP) has published diagnostic criteria for the diagnosis of CRPS types one and two[31] and revisions have already been suggested.[32] Whether the revised clinical and research criteria, or the original criteria are used, objective signs of vasomotor instability (changes in skin blood flow or evidence of temperature asymmetry) remain a diagnostic criterion. This is important as it is well established that palpation alone is a poor way to assess for skin temperature change.

In addition to being insensitive, palpation provides no ability to map the distribution of those changes. In cases where allodynia, hyperalgesia or barometric weather sensitivity exist, only thermography offers the ability to objectify if the vasomotor instability criterion is satisfied. In addition, the American Medical Association's "Guides to the Evaluation of Permanent Impairment" recognizes that "…regional sympathetic blockade has no role in the diagnosis of CRPS." Instead, it cites objective criteria inclusive of vasomotor change.[2]

Perplexed by the CRPS/RSD patient, "The Guides" suggest rating impairment based on alteration in activities of daily living, loss in motion of each joint involved, sensory and motor deficits for the nerve involved or sensory deficits, loss of power, and pain for the body part involved. While this approach attempts to sidestep the problem of objectifying which body part is involved, it is still left to the physician to address the issue.

In this light, the body part involved becomes not only a clinical issue, but also a medical-legal one. Just as a post-stroke, shoulder-hand syndrome patient may present with a swollen hand and be unable to communicate that there is proximal pain, a patient with a crush injury to the hand cannot be expected to articulate that he has vasomotor instability as far proximal as the shoulder or that the perception of compensatory proximal pain is actually sympathetic involvement of the entire limb.

Only after vasomotor mapping has been completed can the distribution of asymmetry be fully determined and the question of which body part is involved be properly addressed. There are many other difficult situations in which thermography is extremely useful in objectifying the extent or presence of involvement. These include thoracic outlet syndrome, cervical-brachial syndrome, vasomotor headache, atypical facial pain, the posterior cervical sympathetic syndrome of Barre-Leiou, and failed back syndrome.[3,11,21]

While a sympathetic component should be considered in each of the aforementioned conditions, TOS deserves special attention. Patients who suffer from this malady often undergo extensive workups only to find the results to be negative. X-ray examination for a cervical rib is only found in a minority of cases and, when present, an even smaller number of cases show positive arteriograms.[7,15]

Although other conditions that may be confused with TOS, such as radiculopathy or ulnar neuropathy, may be exposed, electrodiagnostic studies are usually not diagnostic for TOS.[17] The vast majority of TOS cases are not due to overt vascular or lower trunk neurologic pathology, but rather secondary to numerous musculoskeletal conditions such

Thermography is no different than any other objective study. Ultimately, it is always best to treat patients based upon both clinical and diagnostic impressions, not test results alone.

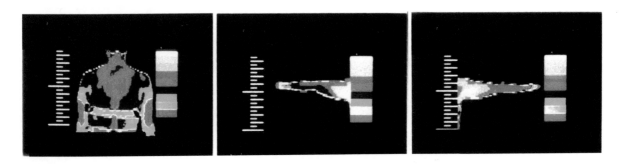

Figure 5: *Heat emission asymmetry pattern involving the medial aspect of the arm and forearm consistent with thoracic outlet syndrome. The rest of the areas showed a symmetric heat emission pattern.*

as scalene anticus spasm, scapulothoracic dysfunction with resultant tension or mechanical torque across the thoracic outlet, and cervical-thoracic interspinous ligament strain with reflex myotomal spasms or myosfasical pain.

Irrespective of which of these somatic issues is the source, patients complain of cold, burning, numb sensations, typically radiating from the neck or shoulder down the medial aspect of the arm and into the fourth and fifth fingers. In some cases, vasomotor instability is visible to the naked eye with obvious skin color changes. These patients frequently respond to a stellate or cervical plexus block.[16]

Thermography is the obvious test of choice to objectify the presence or absence of a vasomotor instability consistent with TOS. Many surgeons have an aversion toward operating upon the TOS patient. In the absence of an absolute surgical indication for TOS, such as an obvious cervical rib that creates clear-cut stenosis on arteriography, thermography is the most cost-effective and diagnostic approach.

A positive study clearly demonstrates a heat emission asymmetry pattern across the medial aspect of the arm and forearm (Figure 5). Radiation to the medial aspect of the fourth and fifth fingers may be present as well. If a study is positive, and the clinician feels the need, then an apical chest X-ray to assess for cervical rib or arteriography can always be obtained later on.

Thermography is ideally suited for diagnosing Barre-Lieou, another common condition.[9] There is no other diagnostic study that can objectify the presence of associated vasomotor instability. In Barre-Lieou, the posterior cervical sympathetic chain generates aberrant impulses that can result in facial heat emission asymmetry patterns (Figure 6).

There are several possibilities as to why the syndrome occurs, including a direct traction injury on the chain as in a whiplash-type injury, ischemia, or hidden infection. In any event, the result is recalcitrant

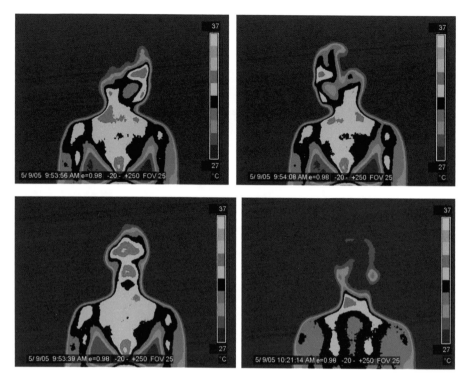

Figure 6: *There is a cold heat emission asymmetry pattern on the maxillary portion of the face and localized hot spots over the omohyoid and nuchal ligament, consistent with Barre-Lieou.*

head and neck pain — with or without scapulo-thoracic pain — associated with blurred vision, tinnitus, vertigo, or nausea.

Barre-Lieou is frequently responsive to sympathetic block. Infrared thermographic imaging of the face, cervical spine, and extremities effectively demonstrates vasomotor asymmetry in these cases. Through its unique mapping ability, thermography can also provide the physician with insight into which somatic level is responsible for the abnormality.

While proving the presence of heat emission asymmetry has great clinical significance, the benefit of objectifying its absence should not be overlooked. When criteria for CRPS are satisfied, but there is no vasomotor abnormality as with RSD, sympathetic independent pain should be more seriously considered.[24] In this instance, relief from sympathetic block is far less likely and alternate conditions or interventions should be considered.

Secretan's Syndrome, which consists of post-traumatic peritendonous fibrosis, brawny edema, loss of finger extensor function, and trophic skin changes, is a relatively uncommon disorder that can mimic CRPS/RSD. This condition has no vasomotor or sudomotor component, so Infrared Thermographic imaging will be negative.

Glomus tumor of the hand, due to neuro-myoarterial tumor formation, is associated with excruciating distal finger pain, cold intolerance, and pain triggered by palpation. Abnormal blood flow in the distal phalanx does occur, but typical vasomotor asymmetry patterns do not.[30]

If it were not for the unique qualities of medical thermography, the information obtained by it would not be otherwise available. Failure to consider the objective information made available by thermography limits clinical assessment and rational decision-making when developing a treatment approach for relevant conditions. Thermography is a unique imaging study that provides the physician with invaluable information in the diagnosis, treatment, and management of patients with suspected or bonafide sympathetic pain syndromes.

References

1. American Academy of Thermology, Practice Guidelines Committee. *Guidelines for Neuromusculoskeletal Thermography,* Wheeling, WV, 2006.

2. American Medical Association. *Guides to the Evaluation of Permanent Impairment,* American Medical Association, Chicago, 2000.

3. Bersani F, Ed. *Electricity and Magnetism in Biology and Medicine,* Kluwer Academic/Plenum Publishers, New York, 1999.

4. Bonica JJ. *The Management of Pain,* Second Edition, Lea and Febiger, Philadelphia, 1990.

5. Borenstein DG, Wiesel SW, and Boden SD, Eds. *Low Back and Neck Pain,* Third Edition, W. B. Saunders, Phildelphia, 2004.

6. Cousins MJ, Bridenbaugh PO, Eds. *Neural Blockade in Clinical Anesthesia and Management of Pain,* Second Edition, Lippincott, Philadelphia, 1998.

7. Cyriax J. *Textbook Of Orthopedic Medicine,* Bailliere Tindall, Philadelphia, 1982.

8. Friedman MS. The Use of Thermography in Sympathetically Maintained Pain. *Iowa Orthop J,* 1994;14:141-7.

9. Gayral L and Neuwirth E. Oto-Neuro-Opthalmologic Manifestations of Cervical Origin – Pos terior Cervical Sympathetic Syndrome of Barre-Lieou. *N Y State J Med,* 1954;1920-6.

10. Gonzalez, E. *The Nonsurgical Management of Acute Low Back Pain,* Demos Vermande, New York, 1997.

11. Govindan S. Infrared Imaging of Extracranial Microcirculation: A Review. *Thermology Int ,* 2003;13:91-98.

12. Gray H. *Gray's Anatomy,* S. Standring, Editor-in-Chief. Elsevier Churchill Livingstone, New York, 2004.

13. Gulevich SJ, Conwell TD, Lane J, et al. Stress Infrared Telethermography is useful in the diagnosis of complex regional pain syndrome, type 1 (formerly reflex sympathetic dystrophy). *Clin J Pain,* 1997;13:50-59.

14. Hooshmand H. *Chronic Pain: Reflex Sympathetic Dystrophy Prevention and Management,* CRC Press, Boca Raton, Florida, 1993.

15. Klippel JH and Dieppe PA, Eds. *Practical Rheumatology.* Mosby, Baltimore, 1995.

16. Kofoed H. Thoracic Outlet Syndrome: Diagnostic Evaluation by Analgesic Cervical Disk Puncture. *Clin Orthop Relat Res,* 1981;156:145-48.

17. LaBan MM, Ed. Physiatric Pearls, *Phys Med Rehabil Clin N Am,* 1996;7:3.

18. Matsumoto S. Thermographic Assessments of Sympathetic Blockade by Stellate Ganglion Block, *Jpn J Anesthes (Masui),* 1991;40(5):692-701.

19. Netter FH. *Atlas of Human Anatomy,* Third Edition, Elsevier Health Sciences, West Cadwell, NJ, 2002.

20. Ochoa JL. The Human Sensory Unit and Pain: New Concepts, Syndromes and Tests, *Muscle Nerve,* 1993;16(10):1009-16.

21. Palmer J, Uematsu S, Jankel W, and Perry A. A Cellist With Arm Pain: Thermal Asymmetry in Scalene Anticus Syndrome. *Arch Phys Med Rehabil,* 1991;72:237-42.

22. Pećina MM, Krmpotić-Nemanić J, and Markiewitz AD. *Tunnel Syndr,* CRC Press, New York, 2001.

23. Raj PP, Ed. *Pain Medicine: A Comprehensive Review,* Second Edition. Mosby, St. Louis, 2003.

24. Roberts W. Adrenergic Mediation of Sympathetically Mediated Pain Via Nociceptive or Non-Nociceptive Afferents or Both? *APS J,* 1992;1(1):12-15.

25. Schwartz R. Electric Sympathetic Block: Current Theoretical Concepts and Clinical Results. *J Back Musculoskelet Rehabil,* 1998;10:31-46.

26. Schwartz, RG. *Resolving Complex Pain,* Color Edition, Piedmont Physical Medicine & Rehabilitation, Greenville, SC, 2006, p. 104.

27. Schwartz, RG. *Resolving Complex Pain,* Color Edition, Piedmont Physical Medicine & Rehabilitation, Greenville, SC, 2006, p. 97.

28. Schwartz, RG. *Resolving Complex Pain,* Color Edition, Piedmont Physical Medicine & Rehabilitation, Greenville, SC, 2006, pp. 223-30, 233-6, 245-6.

29. Taylor GI and Palmer JH. The Vascular Territories (Angiosomes) of the Body: Experimental Study and Clinical Applications. *Br J Plas Surg,* 1987;40:113-41.

30. Waldman SD, Ed. *Atlas of Uncommon Pain Syndromes,* W. B. Saunders, Philadelphia, 2003.

31. Wasner G, Schattschneider J, and Baron R. Skin Temperature Side Differences – A Diagnostic Tool For CRPS. *Pain,* 2002;98:19-26.

32. Wilson PR, Stanton-Hicks M, and Harden RN, Eds. *CRPS: Current Diagnosis and Therapy,* IASP Press, Seattle, 2005.

The Role of Thermography in the Diagnosis and Management of Complex Regional Pain Syndrome

BRYAN J. O'YOUNG, MD
JEFFREY M. COHEN, MD

During the last decade, there has been an expanding role for the use of the thermography (Computerized Infrared Imaging or CII) in the diagnosis and management of Complex Regional Pain Syndrome (CRPS). As pain management becomes an important part of the clinician's role and as CRPS has often been an elusive diagnosis, there is an increased recognition and appreciation of the value of thermography.

This chapter will review the important role of thermography for the diagnosis of patients with CRPS and in the facilitation of its treatment. The chapter begins with the discussion of general principles relative to CRPS and its proper diagnosis. This is followed by a review of the common diagnostic tools used in confirming CRPS. The chapter concludes with a series of case studies reviewing the role of thermography in diagnosing and facilitating the management of CRPS.

Introduction

CRPS is a syndrome characterized by continuous and severe pain in a region of the body without nerve injury (CRPS I or reflex sympathetic dystrophy (RSD)) or with obvious nerve lesions (CRPS II or causalgia) in which the pain is out of proportion to the inciting event.

Causalgia

In 1864, Mitchell, a union army surgeon, introduced the term causalgia, which means burning pain, to describe the persistent symptoms in soldiers following gunshot wounds to peripheral nerves during the American Civil War.[26] The burning pain was often accompanied by additional features including: various sensory disturbances; temperature and sweating changes; glossy and other disturbances of the subcutaneous tissues, muscles and joints; paralysis; and involuntary movements.

Reflex Sympathetic Dystrophy

By the late 1800s, Wolff, Kummel, Sudeck and others noted that there were patients with a similar but less severe condition that resembled causalgia and again often followed trauma but without major nerve injury (without clarification of the term major).

This condition had many synonyms including minor causalgia, posttraumatic vasomotor disorder, posttraumatic osteoporosis, Sudeck's atrophy, algodystrophy, sympathalgia, algodystrophy, shoulder hand syndrome, and reflex sympathetic dystrophy. The last term was coined by Evans in 1946.[15]

The different terms for reflex sympathetic dystrophy have caused great confusion as there has been no single diagnostic criterion that was widely accepted by clinicians and researchers worldwide. The lack of consensus and the confusion with regard to RSD and causalgia made it difficult to conduct research and there was little progress with regard to understanding the pathophysiology of RSD or causalgia.

Complex Regional Pain Syndrome

Diagnosis

In 1993, a consensus group of pain medicine experts from the Special Consensus Workshop of the International Association for the Study of Pain agreed to dismantle the term RSD and causalgia. They decided to rename these syndromes as CRPS[25] with the following definition:

CRPS is a syndrome characterized by continuous and severe pain in a region of the body, usually a limb. The current (1993) diagnostic criteria are:

- the presence of an initiating noxious event, or a cause of immobilization (*optional*)
- continuing pain, allodynia, or hyperalgesia with pain disproportionate to any inciting event

• evidence of edema *at some time*, changes in skin blood flow, or abnormal sweating in the region of pain

This diagnosis is excluded by the existence of conditions that would otherwise account for the degree of pain and dysfunction.

CRPS is classified as Type I without a specific nerve lesion (previously known as RSD) or Type II with an identifiable nerve lesion (previously known as causalgia).

Other features not covered in the criteria may include weakness, dystonia, often with contracture, increased nail growth, increased or decreased hair growth, and osteoporosis (in advanced stages).

Pathophysiology

The precise pathophysiology of CRPS is unknown. There are many proposed mechanisms that include the peripheral somatic peripheral nervous system, the central nervous system (including the spinal cord, brain, and psychological processes), the autonomic nervous system (including both peripheral and central components), and myofascial system. The most current theories can be summarized into two basic categories.

PERIPHERAL SENSITIZATION — Tissue injury causes the release of numerous inflammatory mediators into the tissue. Nociceptive nerve endings in the injured tissue become hypersensitized to mechanical and chemical stimuli. The sensitized nerves have a lower threshold for depolarization and may discharge spontaneously. Consequently, there is increased afferent nerve firing and an augmented experience of pain. In sympathetically maintained pain (SMP), circulating catecholamines may augment this activity; this may be one reason for the efficacy of sympathetic blocks, which reduce catecholamine levels in the blocked limb.

CENTRAL SENSITIZATION — The ongoing "nociceptive barrage" from hypersensitized pain afferents causes chemical and structural changes in the dorsal horn of the spinal cord. *Wide-dynamic-range neurons* in the dorsal horn develop a lower firing threshold and respond to a broader range of stimuli. As a result, the painful area becomes larger

CRPS is a syndrome characterized by continuous and severe pain in a region of the body, usually a limb.

and more sensitive, leading to hyperalgesia and allodynia. The hyper-sensitized segment of the dorsal horn results in increased activities of the anterior and lateral horn cells in the corresponding segment, which in turn produces increased motor activities including increased muscle spasm, increased vasospasm and hyperhydrosis.[1,18]

Most likely, CRPS develops and is maintained by abnormalities in the peripheral nervous system, central nervous system, and autonomic nervous system. In some patients, myofascial dysfunction also plays a role in the development in CRPS. Moreover, psychological factors also alter the entire nervous system and musculoskeletal system during and after the development of CRPS. Thus, CRPS most likely arises not from a single pathophysiologic event; rather, it is a heterogenous disorder that has multiple underlying pathophyisologic events that result in similar signs and symptoms.

Clinical Features

The initial and primary complaint is described as severe, constant, burning, and/or deep aching pain, usually in a limb. Allodynia is usually present; hyperpathia and summation may be present. Any stimulation of the skin is perceived as excruciatingly painful. The pain is most often diffuse and nondermatomal. Initially localized to the site of initial injury, the symptoms tend to become more diffuse with time, usually spreading distal to proximal. Other symptoms may include:

- swelling
- limited mobility from stiffness, weakness, spasm, and/or dystonia
- cold, blue, sweaty skin (some patients start with warm, red, dry skin first)
- muscle cramps
- myofascial pain in other areas (a result of chronic guarding and limited motion)
- hair coarsening and/or loss
- faster or slower nail growth; brittle, cracked, and grooved nails
- tendinitis, contracture, and tissue atrophy
- periarticular osteoporosis
- a small percentage of patients develop "spread" of CRPS to other limbs[17]

CRPS most likely arises not from a single pathophysiologic event; rather, it is a heterogenous disorder that has multiple underlying pathophyisologic events that result in similar signs and symptoms.

Staging

The classical clinical staging of CRPS[3] is:

1. *Acute,* with localized pain, hypersensitivity, and muscle cramps
2. *Dystrophic,* with more diffuse pain and swelling, plus hair and skin changes
3. *Atrophic,* with muscle wasting and a nonfunctional limb.

However, the clinical staging is falling out of favor as clinical experience and recent research argue against its validity. The symptoms and clinical course vary so widely among patients that staging is not very helpful for diagnosis or treatment.[27] In addition, a staging system may be counterproductive in the clinical setting, as it may convey a negative outlook in the later stages and add to the patient's perception of hopelessness when such negativity is not based on scientific study. Patients in the later stages have been observed to make significant clinical improvement with proper long term multidisciplinary treatment.[3]

CRPS, SMP and SIP

The pain in CRPS is often categorized as **sympathetically maintained pain (SMP)** or **sympathetically independent pain (SIP).** It is essential to understand the difference between SMP and SIP.

Sympathetically maintained pain (SMP) is pain maintained by sympathetic efferent innervation or by circulating catecholamines. SMP is diagnosed by performing a sympathetic block, either by anesthetizing the local sympathetic ganglion or infusing sympatholytic medications (phentolamine, guanethidine, and others). If sympathetic block produces significant pain relief, the condition can be categorized as SMP. If the block produces no relief, the condition is categorized as SIP. However, sympathetic blocks have systemic effects and the injectate can anesthetize nerve root and peripheral nerve fibers; therefore, a positive response needs to be interpreted cautiously.

CRPS is a clinical diagnosis, whereas SMP refers to pathophysiology. A patient with CRPS may or may not have SMP. CRPS patients with SMP or SIP may be clinically identical. Besides CRPS, SMP may contribute to other pain syndromes, including peripheral neuropathies, post-herpetic neuralgia, and phantom limb pain. It is also helpful to note that if the CRPS patient primarily has SIP, it does not mean that the sympathetic nervous system is not involved. The criteria for the diagnosis include allodynia or hyperalgesia out of proportion to the inciting event and changes in skin blood flow or abnormal sweating.

Diagnostic Studies

As the diagnostic criteria suggest, CRPS is diagnosed by *history and physical examination*. There are no specific laboratory abnormalities in diagnosing CRPS. X-rays, triple-phase bone scan, sympathetic block, thermography (CII), and electromyography/nerve conduction studies may be helpful, but cannot prove or disprove the diagnosis.[1]

Radiological Testing: Radiographs/MRI/CT

Plain radiograph findings of bony demineralization have been noted in patients with CRPS and this finding may support the diagnosis of CRPS.[33] Although these findings may be helpful, they are not specific to CRPS. Most patients do not demonstrate this abnormality. In addition, demineralization may occur solely from limb disuse and is not specific to CRPS.

MRI/CT scan studies are helpful in demonstrating disc pathology or nerve root compression. However, the pain from CRPS arises primarily from small unmyelinated C fibers and the nerve injury cannot be detected on anatomical radiological testing including MRI, CT scan, and plain radiographs.

Bone Scan

Triple phase bone scan is an objective test that documents the physiological effects of CRPS. Bone scan changes that are described to be consistent with CRPS include distinctive patterns of radiotracer uptake, particularly in the delayed phase with increased uptake in the periarticular regions. Kozin and colleagues reported a sensitivity of 67%, specificity of 86%, and a predictive value of 86% in one study[23] and Davidoff and colleagues reported a sensitivity of 44%, specificity of 92% and a positive predictive value of 61%.[14] The latter retrospective study demonstrated that only 21% of patients with CRPS had bone scan abnormalities consistent with CRPS.[14] Based on the above and other studies, Galer and colleagues concluded that the clinical utility of bone scan in CRPS has not been demonstrated.[3] A comprehensive review of the literature by Lee and Weeks also demonstrated that bone scan is diagnostic in only 55% of RSD patients.[24]

Electrodiagnostic studies

EMG/NCV testing can be used to confirm the presence of large fiber peripheral nerve injury for the diagnosis of CRPS. They are not useful in assessing small fiber abnormalities in sympathetic dysfunction. It is unclear whether the nerve injury documented by the EMG or NCV has any prognostic or therapeutic relevance with regard to CRPS.[3]

Autonomic testing

Quantitative sudomotor axonal reflex can demonstrate autonomic abnormalities. One retrospective study reported that resting sweating output was noted to have diagnostic specificity for RSD.[7] Although potentially useful in research, the clinical utility of these devices has not been established. In addition, these testing devices are not readily available in the clinical setting, and therefore they have limited practical applicability.

Thermography (Computerized Infrared Imaging)

Thermography can assist in the documentation of abnormal skin temperature, particularly in the painful region of CRPS.

Physiology

Thermography has been used effectively as an objective, non-invasive test in the assessment of pain.[16,20,21,22] It records temperature radiating off the skin surface. Cutaneous temperature at rest is largely controlled by the sympathetic nervous system (vasoconstrictor nerves) which closely parallel the somatic sensory nerve distribution.[4] Therefore, when certain pain syndromes are present that affect the sympathetic nervous system, changes in cutaneous blood flow and temperature may be present, reflecting the physiologic response to pain. Sato and Schmidt investigated the reflex relationship between the somatic and sympathetic nervous systems and reported that stimulation of small nociceptive fibers increased the sympathetic firing rate at both the spinal and supraspinal levels.[32] They reasoned that if afferent stimulation is common to both pain perception and temperature control, increased firing from a compressed spinal root would cause increased pain and increased sympathetic vasoconstriction, resulting in decreased temperature along a dermatomal distribution. Thermography has been found to be a useful tool for the objective documentation of sensory and sympathetic dysfunction in peripheral nerves with cutaneous projections. Malfunctioning areas can be reliably demonstrated by thermographic imaging of skin temperature changes.

Cutaneous temperature is largely affected by muscle activity and blood flow. Because thermography records the temperature that is radiating off of the skin surface, the interpretation of these results depends on various environmental, physical and physiological factors that may influence the temperature of the skin. Changes in cutaneous temperature due to environment and muscle activity are minimized in the laboratory by controlling the temperature of the room and having the patients equilibrate to the room temperature without movement for 15 minutes prior to capturing the infrared image.

Criteria for Significance

Identifying areas of pain with thermography has traditionally been performed through the comparison of one side of the body to its corresponding site on the contralateral side, using the side without pain as the "control." Body areas are then determined as being symmetric or asymmetric to each other with regards to temperature. Studies have shown temperature symmetry to be well conserved in homologous areas in the absence of pain.[35, 36] Uematsu S. studied 32 normal subjects and 30 patients with peripheral nerve impairment who ranged in age from 12 to 65 years. Uematsu found that in normal persons the skin temperature difference between contralateral sides of the body was only 0.24 +/- 0.073 °C.[34] He noted that skin temperature differences between corresponding sites on one side of the body compared to the other were not only extremely small, but also very stable throughout the body. Uematsu concluded that there is minimal temperature variation between corresponding sites on different sides of an individual's body.

Studies using thermography have differed on the level of asymmetry considered pathognomonic for RSD/CRPS-I. These studies looked at data comparing mean temperature asymmetry values in normal subjects, non-RSD/CRPS-I medical patients, and RSD/CRPS-I patients. Their asymmetry cutoff values ranged from 0.5°C to 1.0 °C.[30,34,37]

Bruehl et al. addressed the issue of how useful thermography was in the diagnosis of RSD/CRPS-I.[6] They studied 22 RSD/CRPS-I patients and 24 non-RSD/CRPS-I limb-pain patients. They found that a temperature asymmetry of 0.6 °C appeared to be the best for classifying patients as RSD/CRPS-I versus non-RSD/CRPS-I accurately. They felt that in combination with other signs and symptoms typically associated with RSD/CRPS-I (e.g., allodynia, sudomotor changes), the interpretation of thermogram results using the cutoff of 0.6 °C helped increase

Because thermography records the temperature that is radiating off of the skin surface, the interpretation of these results depends on various environmental, physical and physiological factors that may influence the temperature of the skin.

one's confidence in the diagnosis of RSD/CRPS-I. This cutoff could help establish a diagnosis in borderline cases in which atypical sets of RSD/CRPS-I symptoms were noted. They also found that if the clinical concern was more focused on ruling out RSD/CRPS-I, an asymmetry cutoff of 0.8 °C or 1.0 °C was preferable, given the greater specificity observed for these larger asymmetries. In agreement with Bruehl's data and to eliminate false positives, an asymmetry cutoff of 1.0 °C is used in the Kathryn Walter Stein Chronic Pain Laboratory.

There are multiple studies assessing the role of thermography in CRPS. Sensitivity and specificity of thermography has been reported within the 80 percentile range.[38]

Clinical applications of thermography in CRPS

If CRPS is a clinical diagnosis, why is it necessary to perform thermography?

- To provide some form of objective evidence to assure the practitioner of the diagnosis and in initiating a multidisciplinary therapy
- To provide a colorful way for patients to see that they have a pathology that can be objectively documented
- To objectively monitor a treatment response, particularly if the patient has difficulty expressing himself
- To objectively evaluate patients with psychological overtones. In CRPS, the disease is invariably accompanied by severe emotional and psychological abnormalities. As a result, the condition can be mistaken for anxiety, malingering, or hysteria.

The importance of early diagnosis of CRPS

Delays in diagnosis can lead to a loss of functional use of the involved extremity, which could be minimized by early entry into a rehabilitation program.

Thermography Case Studies [10,13,28,29,31,39]

The following case studies demonstrate:
- the role of thermography as an objective tool in confirming the suspician of CRPS (Cases 1–3)
- the role of thermography as an objective tool in evaluating a patient with chronic pain in response to a treatment regimen (Case 4)
- the role of thermography as an objective tool in evaluating a patient with chronic pain who had psychological overtones (Case 5)

Confirming suspicion of CRPS

Case 1

32-year-old man presented with persistent low back pain radiating into his right lower extremity that began after an MVA 14 years ago. The patient's gait had declined due to his pain. Examination revealed diffuse right lower extremity muscle atrophy. X-rays, NCV/EMG, and bone scan were all inconclusive. Physical therapy was of limited benefit in relieving his pain.

Thermography of the lower extremities was performed. The thermograms and the results of the temperature differences are displayed below and on the next page (Figures 1a–d).

Thermography revealed significant asymmetries (pain-associated side cooler) in the anterior thighs, knees, legs and ankles and dorsal feet. Findings were consistent with RSD/CRPS-I involving the right lower extremity.

Figure 1a: *Thermogram, bilateral lower extremities, anterior view*

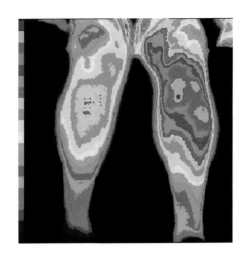

Body Area	Right °C	Left °C	Delta °C
Anterior thigh	31.97	33.24	1.27
Anterior knee	30.21	32.37	2.16
Anterior leg	30.96	33.03	2.07
Anterior ankle	27.37	28.66	1.29

Figure 1b: *Thermogram, bilateral lower extremities, dorsal foot view*

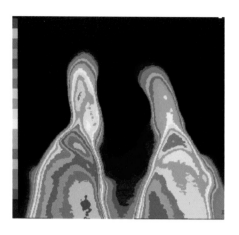

Body Area	Right °C	Left °C	Delta °C
Dorsal foot	26.93	28.51	1.58

Figure 1c: *Thermogram, bilateral lower extremities, posterior view*

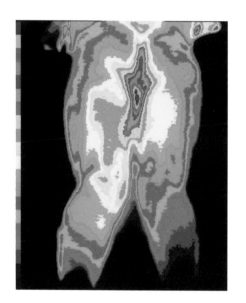

Body Area	Right °C	Left °C	Delta °C
Posterior knee	32.28	33.58	1.20
Calf	30.91	33.01	2.10

Figure 1d: *Thermogram, bilateral lower extremities, sole view*

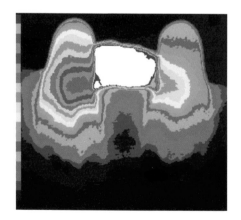

Body Area	Right °C	Left °C	Delta °C
Sole	24.89	27.02	2.13

Case 2

22-year-old man presented with severe right lower extremity pain for over one year which began after he fell into a five-foot pit and injured his right knee. Physical examination revealed increased right lower extremity warmth as well as erythema and edema. Sensory examination was notable for allodynia of his right knee and leg. An MRI of the right knee revealed a sprain of the anterior cruciate ligament. A triple phase bone scan revealed a small area of increased uptake over the right tibial tuberosity with no evidence of RSD/CRPS-I. Patient's gait quality deteriorated as a result of his pain and he required crutches to ambulate. Physical therapy provided only minimal benefit.

Thermography of the bilateral lower extremities was performed. The thermograms and the results of the temperature differences are displayed in Figures 2a–c.

Significant temperature asymmetry was noted (pain-associated side warmer) in his anterior and posterior thighs, knees and legs, posterior ankles, dorsal feet and soles. Findings were consistent with RSD/CRPS-I. Of note, there was a 13.5 °C asymmetry in his bilateral soles.

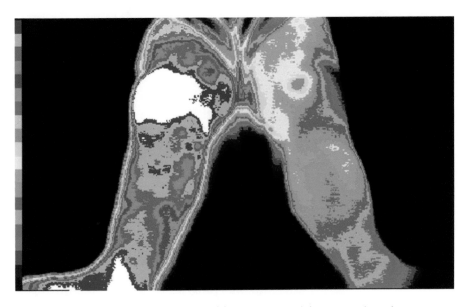

Figure 2a: *Thermogram, bilateral lower extremities, anterior view*

Body Areas	Right °C	Left °C	Delta °C
Anterior Thigh	35.37	31.43	3.94
Anterior Knee	37.52	30.57	6.95
Anterior Leg	35.49	30.59	4.90

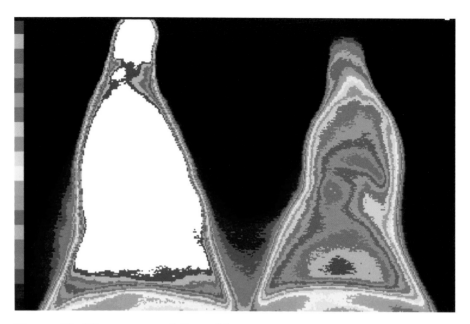

Figure 2b: *Thermogram, bilateral lower extremities, posterior view*

Body Areas	Right °C	Left °C	Delta °C
Posterior Thigh	36.06	32.44	3.62
Posterior Knee	37.49	30.89	6.60
Calf	35.27	29.27	6.00
Posterior Ankle	35.51	26.66	8.85

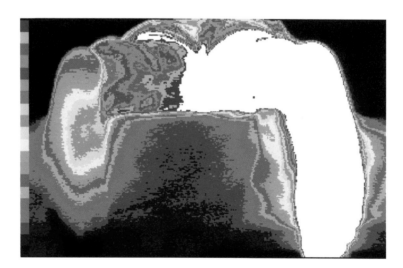

Figure 2c: *Thermogram, bilateral lower extremities, sole view*

Body Area	Right °C	Left °C	Delta °C
Sole	38.38	24.88	13.5

Case 3

72-year-old woman presented with left hemiparesis following a right cerebrovascular accident (CVA) eight months previously. She complained of left upper extremity pain that began soon after her CVA. Her pain radiated down her left arm to her hand. She complained of limited ability to lift her left arm and to use it to perform basic functional activities of daily living skills. Examination revealed slight discoloration and coolness of her left arm and hand. X-rays of her left humerus and wrist revealed profound spotty demineralization. Physical and occupational therapy were of limited value in relieving her pain and restoring mobility. She was felt to possibly have RSD/CRPS-I.

Thermography of the bilateral upper extremities was performed. The thermogram and the results of the temperature differences are displayed in Figure 3.

Thermography revealed significant asymmetries (pain-associated side cooler) in the posterior forearms, dorsal wrists, dorsal hands, dorsal second digits and dorsal third digits. These results confirmed the clinical diagnosis of RSD/CRPS-I.

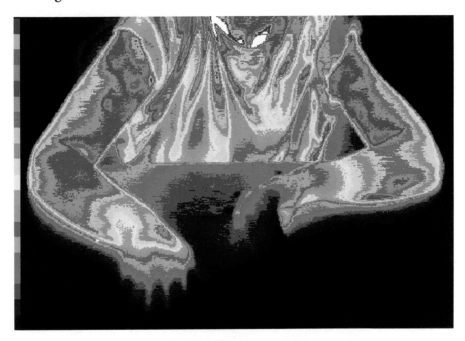

Figure 3: *Thermogram, bilateral upper extremities, anterior view*

Body Area	Right °C	Left °C	Delta °C
Anterior elbow	32.94	31.93	1.01
Anterior forearm	31.34	30.29	1.05
Dorsal hand	30.12	27.04	3.08

Evaluating response to a treatment

Case 4

A 58-year-old female presented with vague complaints of left lower back and bilateral foot and ankle pain for the past 3 months. She had difficulty localizing her pain complaints due to severe dementia. She also reported difficulty walking due to her pain. A comprehensive work-up revealed L34/L45 disc herniation and peripheral polyneuropathy in her lower extremities. On exam, she complained of pain to touch at multiple regions on her back and left lower extremity.

Thermography was performed before and after treatment with transcutaneous electrical nerve stimulation (TENS) and desensitization techniques to the painful regions of her left lower back and left lower extremity. The patient was unable to report her pain status before and after the pain treatment.

Thermography pre-treatment revealed a significant heat asymmetry with her left anterior and posterior ankle significantly warmer than the right (2.90 °C and 3.31 °C, respectively). In addition, her left sole was 3.84 °C warmer than the right pre-treatment. The thermograms and the results of the temperature differences are displayed in Figures 4a-c (on the next page).

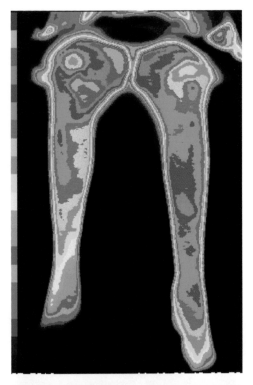

Figure 4a: *Thermogram, bilateral lower extremities, anterior view, pre-treatment*

Body Area	Right °C	Left °C	Delta °C
Anterior ankle	31.89	34.79	2.90

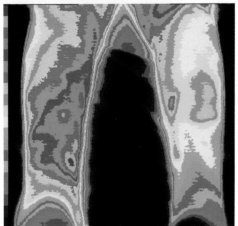

Figure 4b: *Thermogram, bilateral lower extremities, posterior view, pre-treatment*

Body Area	Right °C	Left °C	Delta °C
Posterior ankle	30.76	34.07	3.31

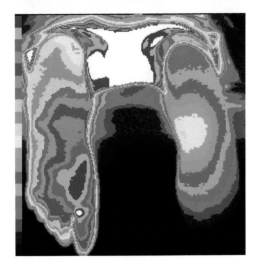

Figure 4c: *Thermogram, bilateral lower extremities, sole view, pre-treatment*

Body Area	Right °C	Left °C	Delta °C
Sole	27.07	30.91	3.84

After two weeks of treatment with TENS and desensitization techniques to her left lower back and left lower extremity, post-treatment images were taken. Her primary medical doctor felt that her condition had improved, but was looking for an objective measurement of improvement in her pain since the patient was unable to communicate her pain levels. Follow-up thermography of her anterior leg revealed less asymmetry at her anterior ankles (left now 1.61 °C warmer than the right) and posterior ankles (left now 2.05 °C warmer than the right). This reflects a decrease in asymmetry of greater than 1 °C in both the anterior and posterior distal lower extremities. In addition, the asymmetry in her soles decreased by 0.65 °C following treatment. Figures 4d–f show the thermograms and the results of the temperature differences.

Figure 4d: *Thermogram, bilateral lower extremities, anterior view, post-treatment*

Body Area	Right °C	Left °C	Delta °C
Anterior Ankle	30.83	32.44	1.61

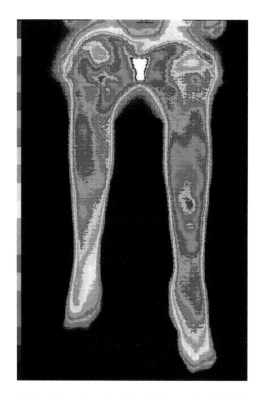

Figure 4e: *Thermogram, bilateral lower extremities, posterior view, post-treatment*

Body Area	Right °C	Left °C	Delta °C
Posterior Ankle	29.89	31.94	2.05

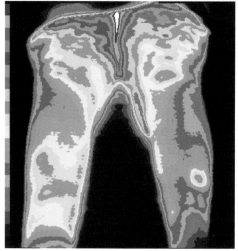

Figure 4f: *Thermogram, bilateral lower extremities, sole view, post-treatment*

Body Area	Right °C	Left °C	Delta °C
Sole	25.68	28.87	3.19

Thermography provided an objective measure of response to treatment (TENS and desensitization techniques) and proved particularly helpful in monitoring the course of treatment for this patient with dementia.

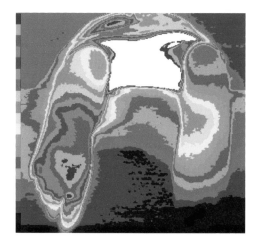

Searching for objective physiological evidence in cases with psychological overtones

Case 5

Patient presented with six months of lower back pain radiating to her right lower extremity. Patient's MRI was negative and EMG was refused. Patient's past medical history was notable for severe borderline personality disorder, self-harming acts, and narcotic abuse.

Thermography of the bilateral feet and ankles was performed. The thermogram and the results of the temperature differences are displayed in Figure 5.

Thermography showed marked asymmetry correlating with pain complaints. Pain substantially decreased with pain cocktails, steroids and interdisciplinary treatments.

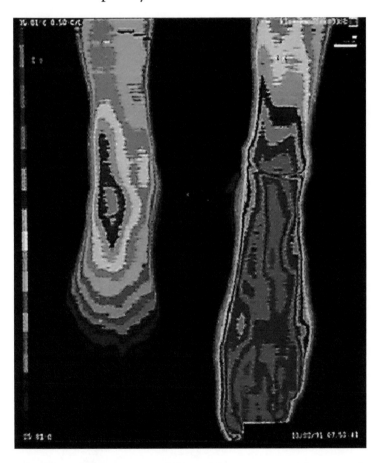

Figure 5: *Thermogram, bilateral lower extremities, anterior view*

Body Area	Right °C	Left °C	Delta °C
Anterior ankle	28.47	32.66	4.19
Dorsal foot	29.61	31.71	2.10

Conclusion

CRPS is a painful and disabling neuropathic disorder. The importance of early diagnosis of CRPS is critical. Delays in diagnosis can lead to a delay in initiating rehabilitation and a possible loss of functional use of the involved extremity. CRPS is a clinical diagnosis and it is challenging to provide objective evidence to confirm its presence. While the traditional diagnostic methods, including X-rays, MRI/CT, bone scan, and EMG/NCS have difficulty confirming the presence of CRPS, thermography provides a safe, noninvasive, and colorful way of confirming CRPS. When used as an adjunctive screening tool, thermography can prove valuable to the medical practitioner not only as a diagnostic tool but also as a therapeutic guide. In addition, the diagnosis of CRPS may not only have a medical relevance, but it can also have legal relevance when the diagnosis is in doubt. Thermography results, when integrated with a thorough clinical evaluation and other relevant testing, can increase one's confidence in the diagnosis of CRPS.

References

1. Birklein F. Complex regional pain syndrome. *J Neurol,* 2005;252:131-8.
2. Bonica JJ. *The Management of Pain,* First Edition, Lea and Febiger, Philadelphia, 1953.
3. Bonica JJ. *The Management of Pain,* Third Edition, Lea and Febiger, Philadelphia, 2000.
4. Brelsford KL and Uematsu S. Thermographic presentation of cutaneous sensory and vasomotor activity in the injured peripheral nerve. *J Neurosurg.* 1985;62:711-5.
5. Bruehl S, Harden RN, Galer BS, et al. Complex regional pain syndrome: are there distinct subtypes and sequential stages of the syndrome? *Pain,* 2002;95:119-24.
6. Bruehl S, Lubenow TR, Nath H, and Ivankovich O. Validation of Thermography in the Diagnosis of Reflex Sympathetic Dystrophy. *Clin J Pain.* 1996;12:316-25.
7. Chelimsky TC, Low PA, Naessens JM, Wilson PR, Amadio PC, O'Brien PC. Value of autonomic testing in reflex sympathetic dystrophy. *Mayo Clin Proc,* 1995;70:1029-40.

8. Cohen JM, O'Young B, Altschul R, and Lee MHM. Computerized Infrared Imaging as an objective assessment tool in evaluating the response to therapeutic interventions for chronic pain. Poster Presentation - International Society of Physical and Rehabilitation Medicine – 3rd World Congress, Sao Paulo Brazil, April 2005.

9. Cohen JM, Wu SSH, Cabrera IN, Haas F, and Lee MHM. The Physiological Documentation of Repetitive Strain Injury using Computerized Infrared Imaging-A Case Series. *Arch Phys Med Rehabil,* 2001;82:1498.

10. Cohen JM, Wu SSH, Yuhn SH, and Lee MHM. Computerized Infrared Imaging in the evaluation of chronic pain in patients in whom the standard diagnostic work-up is negative. *Am J Phys Med Rehabil,* 2003;82:245.

11. Cohen JM, Wu SSH, Yuhn SH, and Lee MHM. Computerized Infrared Imaging as an objective assessment tool in patients undergoing lumbar sympathetic blocks for Complex Regional Pain Syndrome-Type I. *Am J Phys Med Rehabil,* 2003;82:245.

12. Cohen JM, Wu SSH, Yuhn SH, and Lee MHM. Computerized Infrared Imaging as a Tool in Monitoring the Clinical Response to Acupuncture Treatment in a Patient with Chronic Abdominal Pain: A Case Report. *Arch Phys Med Rehabil,* 2003;84:A26.

13. Cohen JM, Yuhn SH, and Lee MHM. The role of Computerized Infrared Imaging as an Objective Assessment tool in diagnosing Complex Regional Pain Syndrome and facilitating its treatment. *Arch Phys Med Rehabil,* 2004;85:E42.

14. Davidoff G, Werner R, Cremer S, Jackson MD, Ventocilla C, and Wolf L. Predictive value of the three phase technetium bone scan in diagnosis of reflex sympathetic dystrophy syndrome. *Arch Phys Med Rehabil,* 1989;70:135-7.

15. Evans JA. Reflex sympathetic dystrophy. *Surg Gynecol Obstet,* 1946;82:36-43.

16. Friedman MS. The Use of Thermography in Sympathetically Maintained Pain. *Iowa Orthop J,* 1994;14:141-7.

17. Harden RN, Bruehl S, Galer BS, et al. Complex regional pain syndrome: are the IASP diagnostic criteria valid and sufficiently comprehensive? *Pain,* 1999; 83:211-9.

18. Harden RN, Rudin NJ, Bruehl S, et al. Elevated systemic catecholamines in complex regional pain syndrome and relationship to psychological factors: a pilot study. *Anesth Analg,* 2004;99:1478-85.

19. Hoosmand, H. Is Thermal Imaging of Use in Pain Management? *Pain Digest,* 1998;8:166-170.

20. Jones BF. A Reappraisal of the Use of Infrared Thermal Image Analysis in Medicine. *IEEE Trans Med Imaging,* 1998;17:1019-27.

21. Karpman HL, Sheppard JJ, Clayton JC, Kalb IM. The Use of Thermography in a Health Care System for Stroke. *Geriatrics,* 1972;27:96-105.

22. Karstetter KW and Sherman RA. Use of Thermography for Initial Detection of Early Reflex Sympathetic Dystrophy. *J Am Pod Med Assoc,* 1991;81:198-205.

23. Kozin F, Soin JS, Ryan LM, et al. Bone scintigraphy in the reflex sympathetic dystrophy syndrome. *Radiology,* 1981;138:437-43.

24. Lee GW and Weeks PM: The role of bone scintigraphy in diagnosing reflex sympathetic dystrophy. *J Hand Surg* [Am], 1995;20:458-63

25. Merskey H, Bodguk N, Eds. *Classification of Chronic Pain, Descriptions of Chronic Pain Syndromes, and Definition of Pain Terms,* 2nd ed., IASP Press, Seattle, 1994.

26. Mitchell SW, Morehouse CR, and Keen WW. *Gunshot wounds and other injuries of the nerves.* J.B. Lippincott, Philadelphia, PA, 1864.

27. O'Young B, Rudin N, Slaten W, and Wang N. "Complex Regional Pain Syndrome: Assessment and Management," In *Physical Medicine and Rehabilitation Secrets,* 3rd edition, O'Young B, Young M, Stiens S, Eds., Elsevier, 2007.

28. O'Young B, Cohen JM, Wu SSH, Altschul R, and Lee MHM. Computerized Infrared Imaging: Its Role in the Diagnosis of Complex Regional Pain Syndrome. Poster Presentation. International Society of Physical and Rehabilitation Medicine – 3rd World Congress, Sao Paulo Brazil, April 2005.

29. O'Young B, Wu SSH, Cabrera IN, Roesch B, and Lee MHM. The Role of Computerized Infrared Imaging in Complex Regional Pain Syndrome. American Academy of PM&R – Poster Presentation. Annual Assembly, San Francisco, November 2000.

30. Pochaczevsky R. Thermography in posttraumatic pain. *Am J Sports Med,* 1987;15(3):243-50.

31. Richter EF, Wu SSH, Cohen JM, Cabrera, IN, and Lee MHM. Computerized Infrared Imaging as a Diagnostic Tool in Shoulder-Hand Syndrome. *Am J Phys Med Rehabil,* 2001;80:318.

32. Sato A and Schmidt RF. Somatosympathetic reflexes: afferent fibers, central pathways, discharge characteristics. *Physiol Rev,* 1973;53:916-47.

33. Tollison CD and Satterwaite JR, Eds. Sympathetic Pain Syndromes: Reflex Sympathetic Dystrophy and Causalgia (State of the art reviews: Physical Medicine and Rehabilitation), June 1996;10(2). Hanley and Belfus Inc., Philadelphia.

34. Uematsu S. Thermographic imaging of cutaneous sensory segments in patients with peripheral nerve injury. *J Neurosurg,* 1985;62:716-20.

35. Uematsu S, Edwin DH, Jankel WR, et al. Quantification of thermal asymmetry Part I: Normal values and reproducibility. *J Neurosurg,* 1988;69:552-5.

36. Uematsu S, Edwin DH, Jankel WR, et al. Quantification of thermal asymmetry Part II: Application in low-back pain and sciatica. *J Neurosurg,* 1988;69:556-61.

37. Uematsu S, Hendler N, Hungerford D, et al. Thermography and Electromyography in the Differential Diagnosis of Chronic Pain Syndromes and Reflex Sympathetic Dystrophy. *Electromyogr Clin Neurophysiol,* 1981;21:165-82.

38. Wexler CE and Chafetz N. Cervical, thoracic, and lumbar thermography in the evaluation of sympathetic workers' compensation patients — a study. Special supplement of Modern Medicine: *Academy of Neuromuscular Thermography Clinical Proceedings,* 1987;55:53-57.

39. Wu SSH, Cohen JM, Richter E, Cabrera IN, and Lee MHM. Role of Infrared Imaging in the diagnosis of Complex Regional Pain Syndrome Type I in Post-CVA patients. *Arch Phys Med Rehabil,* 1999;80:1167.

7

Thermography — Clinical Indications

JEFFREY M. COHEN, MD

This chapter provides a comprehensive overview of the role of thermography in the evaluation of common musculoskeletal, neurological, vascular and dermatological conditions. It is based upon a review of the literature and highlights clinical areas where thermography is felt to be a valuable diagnostic tool and those areas where it is not recommended. (Note: For discussion of Complex Regional Pain Syndrome, see Chapter 6.)

Myofascial Pain Syndrome

One of the most frequent conditions that develops secondary to local injuries or an overload of muscles is myofascial pain syndrome (MPS). The diagnostic hallmarks of myofascial pain are trigger points. Trigger points, in addition to being locally tender, can refer pain to a remote zone. Thermography represents the first objective measure capable of documenting trigger points. Fischer, in 1984, described the typical thermographic picture of a trigger point.[20] He indicated that it consisted of a localized area of temperature elevation — usually 5-10 mm in diameter — and was frequently disc-shaped. He noted that these findings could be corroborated by local palpation or quantitatively measured by pressure threshold over the involved area. Pressure threshold measurement (PTM) is defined as the minimum pressure (force) that induces pain or discomfort when applied over a tender area. As described by Fischer, the PTM can be determined using a force gauge, which is attached to a rubber disc with a surface area of 1 cm^2. Fischer and Chang have shown that under hot spots, the pressure threshold was statistically significantly lower.[19]

Weinstein, at The Center for Neuromuscular Injury, Old Bridge, New Jersey has used thermography to evaluate over 250 patients with pain symptomatology attributed to the cervical region.[43] He has developed an easily performed and reproducible thermographic technique that enables the practitioner to clearly establish the presence or absence of trigger points. His data demonstrate clearly that trigger point and myofascial pathology can be visualized using thermography. His procedure for establishing a positive diagnosis of a trigger point relies upon the clear demonstration of four standards:

1. The anatomical area of concern must be shown to be at least 1 °C above the circum-ambient temperature.
2. The area must match the patient's pain diagram.
3. The area must be evident throughout three series of pictures taken at 15-minute intervals during standard thermography.
4. The area of increased temperature must not be extinguishable by the brief application of alcohol spray to the involved surface.

Hakguder et al. investigated whether low level laser therapy had a clinical therapeutic effect in myofascial pain syndrome (MPS) and used thermographic evaluation as one of their outcome measures.[23] They evaluated 62 patients with MPS who had an active trigger point in the neck or upper back and divided the patients into two equal groups. The first group was treated with low-level laser therapy and stretching exercises and the second group received stretching exercises alone. Outcome measures were pain measured by the visual analogue scale, algometry on the trigger point, and thermal asymmetry. The authors found that mean pain values decreased more significantly in the group receiving the combined therapies from baseline to three weeks follow-up. This clinical response paralleled the thermographic changes in patients who were treated with the additional laser treatment. The authors felt that their findings supported the use of thermography as an evaluation method for the efficacy of laser treatment in MPS.

Low Back Pain/Radiculopathy

Thermography has been proposed by some as a safe, economic and effective diagnostic test for lumbar radiculopathy. It has also been recommended by some for evaluation of low back pain. The use of thermography for diagnosing lumbar radiculopathy was first cited by Albert in 1964.[2] Weinstein reported a 95% accuracy for thermography in documenting the presence of low back pain compared to 85% for CT, 90% for EMG and 66% for myelography.[44] Proponents of thermography claimed that correlating an abnormal thermogram with an anatomic imaging defect was convincing evidence of radiculopathy and that a normal thermogram may obviate the need for further work-up.[10]

So et al. in *Neurology* (1989), used thermography to study 27 normal subjects and 30 patients with low back pain, 21 of whom had clinical signs of lumbosacral radiculopathy.[36] They performed this study to evaluate the accuracy of thermography in diagnosing lumbar radiculopathy and compared its accuracy to conventional electrodiagnostic techniques. After an initial clinical examination, each patient was evaluated by thermography as well as by conventional EMG and nerve conduction studies. The authors found that 17 of the 21 patients (81%) had abnormal thermographic studies, whereas 15 of the 21 (71%) had abnormal electrodiagnostic studies. However, in only three of the patients, did thermography permit the correct level of the lesion (as judged clinically) to be determined. It failed to localize the abnormal root level in 11 of the 12 patients who had single-level radiculopathy on electrodiagnostic testing. A diffuse pattern of thermographic abnormality involving both the dorsal and plantar surfaces of the foot (L5 and S1 territory) accounted for many of the non-localizing findings. The authors also noted that when a thermographic abnormality existed, a significant inter-side difference (exceeding 3 standard deviations from the normal mean) was seen in only a small portion of the dermatome.

The authors concluded that thermography was similar in sensitivity to conventional electrodiagnostic studies. They noted that the presence of a thermographic abnormality correlated well with the presence of clinical and EMG abnormalities. However, they indicated that their study raised serious questions about the localizing value of thermography. Both the level of root involvement and the side of the root lesion could not be identified with certainty on thermographic grounds. The authors felt that although thermography was non-invasive and painless, electrodiagnostic studies were advantageous in providing more specific information about the integrity of the nerve roots. They concluded that although thermography may be abnormal in persons with lumbosacral radiculopathy, the findings were nonspecific and of uncertain prognostic significance. Therefore, the authors could not recommend the routine use of thermography for the diagnosis of lumbosacral radiculopathy.

Hoffman et al. in *Spine* (1991) undertook a literature search and meta-analysis to evaluate thermography's diagnostic accuracy and clinical utility.[25] Using MEDLINE, they searched for English-language clinical studies of thermography and low back pain between 1971 and 1990. Twenty-eight studies were evaluated for diagnostic-accuracy data (sensitivity and specificity) and method. They found that there were major methodological flaws in twenty-seven of the studies, including small sample size, faulty cohort assembly, poor clinical descriptions, and biased test interpretations. Based on their research, they felt that the role of thermography in the diagnosis of lumbar radiculopathy remains unclear. They felt that thermography could not be recommended for routine clinical use in evaluating low back pain and that rigorous clinical research was required to establish thermography's diagnostic accuracy and clinical utility in lumbar radiculopathy.

Repetitive Strain Injury

Repetitive motion of the hands and wrists, as in typing, can result in a repetitive strain injury, with the onset of incapacitating forearm pain. This condition, which may have diverse symptoms, often has few physical signs and a lack of objective measurements to document it.[22] As clinical observations have suggested the presence of vasomotor changes in repetitive strain injury, Sharma et al. embarked on a study using thermography to evaluate this condition.[35] They studied ten consecutive keyboard operators with chronic forearm pain exacerbated by keyboard work and 21 asymptomatic controls matched for sex and typing speed (30-50 words/min). All keyboard operators had diffuse forearm pain for at least three months. A Talytherm infrared camera unit was used and baseline images were taken after acclimatization for five minutes at 22–24 ºC. Subjects then typed at their usual speed for five minutes and another thermogram was taken immediately afterwards. Thermograms were taken over the 2nd, 3rd, and 4th proximal phalanges of both hands, avoiding large muscle masses. The authors found that mean temperatures readings before typing were similar in the two groups. However, all keyboard operators had symptoms following typing and their mean temperature readings were significantly reduced (mean 2.11 ºC, range 0.45-3.44 ºC). Only four controls showed cooling (mean 0.55 ºC, range 0.35-1).

The authors felt that the cooling seen in the keyboard operators may have been due to sympathetic overactivity as a result of nocioceptor and mechanoreceptor stimulation leading to a reflex neuropathic state. They concluded that thermography may have a role in evaluating repetitive strain injury, particularly in measuring response to treatment, but that further research was needed.

Gold et al. in *European Journal of Applied Physiology* (2004) evaluated three groups of office workers by thermography following a 9-minute typing challenge.[21] These groups included office workers with upper extremity musculoskeletal disorders such as tendonitis and carpal tunnel syndrome with or without cold hands exacerbated by keyboard use, and control subjects. Using infrared thermography, they were able to distinguish between the three groups of subjects. The authors felt that the keyboard-induced cold hand symptoms were due, at least partially, to reduced blood flow.

Osteoarthritis

Infrared thermal images provide a non-contact means of measuring surface temperature. Several thermographic studies have been performed to evaluate rheumatic diseases of the hand and osteoarthritis (OA) of the knees and hands. For example, thermography has been used to provide objective data on joint inflammation in response to therapeutic interventions. Dieppe et al. in *Rheumatology Rehabilitation* (2004) used infrared thermography to evaluate temperature changes in knee OA in response to intra-articular steroids.[17] They evaluated twelve patients and found a significant reduction in joint surface temperature one week after intra-articular steroid injection.

Varju et al. in *Rheumatology* (2004) hypothesized that measurement of joint surface temperature could complement the radiographic evaluation of OA by providing dynamic and physiological information to augment the static information from X-rays.[41] They evaluated 91 subjects with bilateral clinical hand OA both thermally and radiographically (Figure 1). Temperature measurements were made on digits 2–5 of both hands at the DIP (distal interphalangeal), PIP (proximal interphalangeal) and MCP (metacarpophalangeal) joints (2184 total joints). The authors found that joint surface temperature varied with the severity

Figure 1: *Side-by-side comparison of thermographic, X-ray, and digital images in right hand osteoarthritis.*

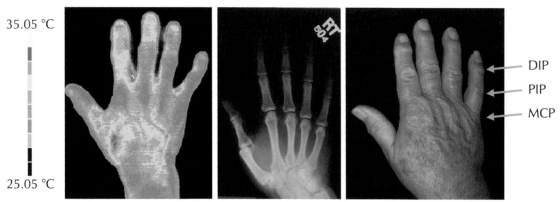

Reprinted by permission of Oxford University Press on behalf of the British Society for Rheumatology.[41]

of radiographic OA. The earliest discernable radiographic disease was associated with higher surface joint temperatures. The authors felt that their results support the notion that the earliest discernable radiographic stage of hand OA represents an inflammatory phase. Joints with increasing severity of radiographic OA were associated with lower surface temperatures than normal joints. Based on their findings, the authors felt that thermography may provide a non-radioactive and non-invasive method of evaluating the presence and severity of OA. In addition, it may provide a sensitive modality to follow the response of OA to disease modifying agents.

Peripheral Nerve Injuries

Thermography is a useful tool for the objective documentation of sensory and sympathetic dysfunction in peripheral nerves. Brelsford and Uematsu in *Journal of Neurosurgery* (1985), used computerized thermography to document temperature changes resulting from the local anesthetic block of peripheral nerves in two Rhesus monkeys.[9] They selected the Rhesus monkey for its anatomical similarity to man, both in its body habitus and peripheral nerve anatomy. One percent xylocaine was injected to block the median nerve at the elbow, the ulnar nerve at the postero-medial elbow, the peroneal nerve at the head of the fibula, and the posterior tibial nerve at the popliteal fossa. Following injection, thermograms were made of the limbs at intervals ranging from 1 to 50 minutes. The thermogram of the contralateral limb served as a control for the affected distribution of the anaesthetized nerve. The authors demonstrated via thermography that there was an increase in skin temperature in the area of the affected sympathetic distribution. They reasoned that by injecting lidocaine, they effectively performed a pharmacological transection of the nerve fiber and thus sympathetic input for vasoconstriction was abolished. This then led to vasodilation and an increase in blood flow and a consequent increase in skin temperature to the area of sympathetic distribution. This increased skin temperature could be displayed both qualitatively and quantitatively via thermography. The authors concluded that thermography was a useful tool for the objective documentation of sensory and sympathetic dysfunction in peripheral nerves.

Uematsu in *Journal of Neurosurgery* (1985) found that thermographic imaging of the cutaneous nerve segment was a clinically useful, sensitive technique that made possible the objective evaluation of what was formerly the patient's subjective expression of sensation.[40] He indicated that the sensory examination, which is based on the patient's subjective assessment of symptoms, can be influenced by many factors and the interpretation of results can often be difficult. As skin temperature is altered in the field of an impaired peripheral nerve due to sympathetic vasomotor dysfunction, Uematsu reasoned that thermography could be used to evaluate the sensory segment in the area of an impaired peripheral nerve. He studied 32 normal subjects and 30 patients with peripheral nerve impairment, who ranged in age from 12 to 65 years. He mapped the skin surface areas to be measured by dividing the body's skin surface into 64 sensory "box" segments (32 per side) that approximated areas of innervation of the peripheral nerves. The skin temperature differences between the sides of the body were determined in the normal subjects and in the patients (Figure 2).

Uematsu found that in normal persons, the skin temperature difference between sides of the body was only 0.24 +/- 0.073 ºC (Table 1). In

Figure 2: *Diagram showing the box segments used in the study for the front and back parts of the body. A total of 64 segments were studied (32 per side). Each box segment covers the distribution of a clinically important nerve section of the body.*

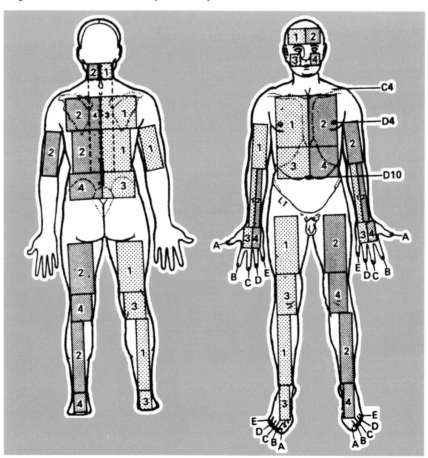

Reprinted with permission of the Journal of Neurosurgery

contrast, in patients with peripheral nerve injury, the temperature of the skin innervated by the damaged nerve was altered an average of 1.55 °C. He further divided the patients with peripheral nerve injury into two groups, based on the status of their sympathetic nerve function. Group A patients had a complete loss of sympathetic nerve function, and a complete loss of sensation in the skin segment relating to the damaged nerve. In these patients, the skin temperature of the damaged side averaged 1.92 degrees higher than the opposite intact limb. Group B patients had a partial transection of a peripheral nerve, which resulted in increased sympathetic nervous activity. In these patients, the area of the damaged nerve segment was typically numb but did not lose complete sensation in all cases. In these patients, the skin temperature in the area of the damaged nerve averaged 0.83 degrees colder than the opposite intact segment.

Uematsu concluded that there is minimal temperature variation between corresponding sites on different sides of an individual's body. Therefore, the detection of a significant temperature difference between corresponding sites on opposite sides of the body is highly suggestive of nerve impairment. He felt that thermography was a useful objective technique that was able to quantify sensory change, and therefore would prove an important tool in evaluating what was formerly the patient's subjective expression of sensation. He also felt that thermography would prove valuable as an aid in the evaluation of disability claims, helping to rule out malingering cases.

Table 1: *Average skin-temperature differences between sides of the body for segments measured.*

Sensory Segment	No. of Cases	Temperature Difference (°C)	Standard Deviation (±°C)
Head			
forehead	29	0.12	0.093
cheek	29	0.18	0.186
Trunk			
chest	11	0.14	0.151
abdomen	11	0.18	0.131
cervical paraspinal area	11	0.15	0.191
thoracic paraspinal area	11	0.15	0.092
lumbar area	10	0.25	0.201
Trunk, average		0.17	0.042
Extremities			
shoulder	10	0.13	0.108
bicpes	10	0.13	0.119
triceps	10	0.22	0.155
forearm			
lateral	19	0.32	0.158
medial	19	0.23	0.198
palm			
lateral	21	0.25	0.166
medial	21	0.23	0.197
thigh			
anterior	14	0.11	0.085
posterior	11	0.15	0.116
knee	14	0.23	0.174
popliteal area	14	0.12	0.101
leg			
anterior	16	0.31	0.277
calf	15	0.13	0.108
foot (top)	15	0.30	0.201
heel	15	0.20	0.220
Extremities, average		0.20	0.073
Fingers, average*		0.38	0.064
Toes, average*		0.50	0.143
* Five segments were measured, but only the average is given here.			

Table reprinted with permission of the Journal of Neurosurgery

Entrapment Neuropathies

The diagnostic reliability of thermography in evaluating patients with focal lesions of peripheral nerves, such as carpal tunnel syndrome at the wrist and ulnar neuropathy at the elbow, has also been studied. Herrick and Herrick in *The Journal of Hand Surgery* (1987) compared the results of thermographic and electophysiologic studies in 55 patients with carpal tunnel syndrome.[24] They reported that the thermographic studies demonstrated a 100% sensitivity and 97% specificity when compared to electrodiagnostic studies. However, there was no control group in their study and the thermographic and clinical criteria for diagnosis were not adequately specified.

So et al. in *Neurology* (1989) studied 20 normal subjects, 22 patients with carpal tunnel syndrome, and 15 patients with ulnar neuropathy at the elbow to compare the diagnostic accuracy of infrared thermography with that of conventional electrodiagnostic studies.[37] The authors used a Bales Scientific MCT 7000 thermography unit. Subjects were studied in an air-conditioned room maintained at a steady temperature of between 20 and 23 °C. Three sets of thermographic images were taken at 15-minute intervals. Each set of images consisted of anterior and posterior views of the upper limbs and trunk, as well as a close-up view of the palmar and dorsal aspects of the hands and digits. Thermography was performed first to avoid the potential effects of the electrodiagnostic examination on skin temperature.

The authors found that thermography was abnormal in 12 of the 22 (55%) patients with carpal tunnel syndrome and seven of the 15 (47%) patients with ulnar neuropathy at the elbow. The abnormalities consisted of an increase in inter-side temperature difference in the fingers and hands (greater than 2.5 SD from the normal mean) or an alteration in the normal thenar-hypothenar temperature gradient in the fingers (first three digits normally warmer than the little finger). The clinical diagnosis was confirmed electrophysiologically in 19 of 22 (86%) patients with carpal tunnel syndrome and 13 of 15 (87%) with ulnar neuropathy. The authors noted that there was little correlation between the severity of the electrophysiologic and thermographic abnormalities. In addition, they noted that the interpretation of inter-side differences could be complicated by the presence of bilateral lesions in entrapment neuropathies.

The authors concluded that the thermographic findings were nonspecific and did not reliably identify the symptomatic side nor did they distinguish between median and ulnar nerve involvement. They indicated, however, that the yield of thermography may be improved by the use of various stress tests, such as hand immersion in cold water or provocative limb postures. They concluded that thermography is not sensitive enough to substitute for conventional electrodiagnostic studies in the evaluation of carpal tunnel syndrome and ulnar neuropathy.

Ming et al. in *Pathophysiology* (2005) sought to clarify sympathetic pathology in carpal tunnel syndrome (CTS) and the usefulness of digital infrared thermography as a diagnostic aid.[29] They reasoned that since sympathetic fibers are very thin and unmyelinated, they would be especially vulnerable to damage in the early phase of carpal tunnel syndrome. This would lead to changes in vasomotor activity in the innervated area with subsequent changes in skin temperature. The authors evaluated 38 clinically diagnosed carpal tunnel hands from 30 patients and 41 hands from 22 healthy volunteers. They found that the temperatures of the median nerve distribution area between the CTS hands and the control group hands were significantly different. In addition, in the CTS hands, the temperatures between the median and ulnar nerve distribution areas were significantly different. The sensitivity and specificity of digital infrared thermography was 84 and 91% respectively. The authors concluded that digital infrared thermography could objectively evaluate sympathetic neural pathology in carpal tunnel syndrome. It could be useful as an additional non-invasive tool in the diagnosis of carpal tunnel syndrome, especially in the early stages.

> The authors concluded that digital infrared thermography could objectively evaluate sympathetic neural pathology in carpal tunnel syndrome.

Vasospastic Disease/Raynauds Phenomenon

Digital vasospasm (Raynaud's syndrome) is a painful, disabling condition that can lead to ulceration and gangrene. Infrared thermography is a useful tool in the diagnosis of vasospastic disorders. It provides an assessment of the microcirculation and its responses to changes in temperature.

Von Bierbrauer et al. in *VASA* (1998) evaluated the use of infrared thermography in the diagnosis of Raynaud's phenomenon of occupational origin, also known as Vibration-Induced White Finger (VWF).[42] This syndrome is due to chronic exposure to vibrating tools such as chain saws or pneumatic hammers. It is felt that this chronic exposure can lead to hypertrophy of the vascular smooth muscle of digital arteries with increased vascular reactivity. Patients develop vasospastic blanching followed by local cyanosis. Vasospastic attacks can be triggered by humidity, cold exposure and vibration. The authors examined 24 patients with VWF using thermography and compared their data to 12 matched healthy controls. Infrared imaging measurements were conducted simultaneously on both hands before, directly after, and 5, 10, 15, 20 and 30 minutes following a 3-minute exposure of both hands to a water bath of 12 °C. Their results indicated that the basal thermograms did not differ significantly between groups. However, the mean finger temperatures after cold exposure were significantly lower in VWF patients compared to controls. The authors found that the parameter of re-warming time after cold exposure was diagnostic: a complete rewarming of the fingers by 15 minutes was indicative of a normal vascular reactivity whereas an incomplete rewarming at 30 minutes following cold provocation was indicative of an impaired vascular reactivity due to VWF. Thus, patients with VWF demonstrated a more intense cooling reaction and significantly delayed re-warming after cold provocation. The authors concluded that infrared imaging was a useful tool in the diagnosis of Raynaud's phenomenon due to cold provocation. It provided an objective assessment of vascular reactivity. In contrast to other methods of objective testing such as finger systolic pressure measurements, sphygmanography, or laser Doppler fluxmetry, it offered the advantage of simultaneous examination of all fingers. This is important in secondary Raynaud's phenomenon in which there may be asymmetric involvement of the digits.

Darton and Black in the *British Journal of Rheumatology* (1991) performed thermographic studies of the hands of 58 patients with primary or secondary Raynaud's phenomenon and of a group of normal subjects.[16] Thermographic images were taken before and after cold provocation. Their results revealed differences in thermal distribution patterns for patients with primary Raynaud's phenomenon (RP), secondary Raynaud's phenomenon (RS), and normal controls. In primary Raynaud's patients, there was a drop in temperature from the palm to the fingertips, with the palm the warmest area in nearly every case and the little finger, the coldest. In patients with secondary Raynauds, the thermal hand pattern was patchier. The authors found that the mean temperature difference between the warmest and coldest areas of the hand was as follows: RP patients ($3.8 + 0.5$ °C), RS patients ($3.9 + 0.2$ °C), normal subjects (less than 3 °C). In both patient groups, the abnormalities in hand pattern and temperature range were exaggerated after cold challenge. The mean temperature range decreased after the cold challenge in the normal subjects, due to reactive hyperemia in the hand. The authors felt that the most notable finding of their study was the patchiness of the hands, noted on thermography, in many patients with RS, which was sometimes revealed only after cold provocation. They feel that this finding has diagnostic potential, particularly if occurring in conjunction with abnormal nailfold capillaries associated with connective tissue disease. They also feel that if hands have a temperature range of more than 3 °C, it may indicate an abnormality.

Clark et al. in *The Journal of Rheumatology* (1999) used infrared thermography to further explore and quantify different thermographic patterns, which distinguish secondary from primary Raynaud's phenomenon.[15] They imaged the hands of 9 patients with primary Raynaud's phenomenon (PRP) and 20 patients with secondary Raynaud's phenomenon (due to systemic sclerosis (SSC)). They used an Inframetrics 600M thermal imaging camera at room temperatures of 23 °C and 30 °C. They then measured the distal-dorsal difference (by subtracting the temperature of the fingertip from the temperature of its corresponding dorsum) for all fingers, excluding the thumb. The authors then performed a cold challenge test at a room temperature of 23 °C by immersing the hand into water at 15 °C for one minute. They then thermally imaged the rewarming process for 15 minutes. The authors found that in patients with SSC, the distal-dorsal difference was higher (colder fingers), both at 23 and 30 °C. They found that the best discriminant between PRP and SSC was the finding of a distal-dorsal difference of greater than 1 °C at 30 °C (the fingers colder than the dorsum by greater than 1 °C), and was specific for underlying connective tissue disease. They noted that having a distal-dorsal difference greater than 1 °C at 23 °C showed borderline discrimination between PRP and SSC.

The authors felt that their findings are physiologically relevant and that the degree of the temperature gradient along the finger relates to the degree of vascular structural obstruction. They noted that in normal subjects, the fingertips were warmer than the dorsum of the hand and that this was reversed in patients with Raynaud's Phenomenon. In patients with PRP, the coldness of the fingers was felt to be due to digital artery spasm and thus the fingers became warmer in a warm room. However, in patients with SSC, the impaired blood flow to the fingers was due to both arterial spasm and to structural vascular disease. Thus, when exposed to a warmer room, although the fingers may warm, one or more of them remain colder than their dorsum (a positive distal-dorsal gradient). The authors concluded that the distal-dorsal temperature difference measured by thermography at two room temperatures enhances the ability to distinguish between primary and secondary Raynaud's phenomenon.

Anderson et al. in *Rheumatology (Oxford)* (2006)[3] found that parameters derived from thermography (incorporating both a heat and cold challenge) aided in the prediction of systemic sclerosis in patients with Raynaud's phenomenon.

Schuhfried et al. in *Archives of Physical Medicine and Rehabilitation* (2000) studied 86 patients referred for the clarification of the diagnosis of secondary Raynaud's phenomenon.[34] They sought to determine the major infrared thermographic parameters to discriminate between patients with and without secondary Raynaud's phenomenon. In each patient, the authors determined the absolute temperature of the fingertips and the longitudinal temperature difference between the fingertips and the radial carpal-metacarpal joint region before, 10, and 20 minutes after a cold challenge. They also calculated the mean area under the re-warming curve of the fingertips, the recovery index 20 minutes after cold challenge, and the most rapid phase of re-warming of the fingertips of both hands. The authors found that the only thermographic parameter to discriminate between patients with and without definite secondary Raynaud's phenomenon was the longitudinal temperature difference between the fingertips and the radial carpal-metacarpal joint region, before cold challenge. The more negative the longitudinal temperature difference, the greater the likelihood of the patient having secondary Raynaud's phenomenon.

Cherkas et al. in *The Journal of Rheumatology* (2003) employed thermographic measurements of digital skin temperature after cold challenge in classifying Raynaud's phenomenon (RP) in a healthy population.[13] They studied 175 patients with RP and 404 controls and subjected them to a 15 °C, 60-second cold challenge test. They found that a low baseline digital skin temperature was a significant predictor of RP and could help to predict the occurrence of RP in patients drawn from the general population. They felt that detecting a low baseline digital temperature may be a useful adjunct to the clinical history in classifying the disease.

Al-Awami et al. in *VASA* (2004) evaluated the clinical and thermographic effects of low level laser irradiation in patients with primary or secondary RP.[1] In a double blind, placebo-controlled study, 47 patients suffering from primary or secondary RP were assigned to receive either 10 sessions of low level laser irradiation or placebo irradiation. The authors found that the frequency and severity of Raynaud's attacks were significantly reduced in patients with primary and secondary RP who had received low level laser irradiation. In addition, the thermographic response to cold challenge improved only in patients treated with low level laser irradiation.

Neuropathic Ulceration/Charcot Arthropathy

Two of the most devastating sequelae of diabetic peripheral neuropathy are neuropathic ulceration and neuropathic osteoarthropathy (Charcot foot). There have been numerous theories regarding the pathogenesis of neuropathic osteoarthropathy. One proposed theory is that following the development of autonomic neuropathy (autosympathectomy), there is a generalized increase in blood flow to an affected limb, resulting in osteopenia that weakens bones of the foot and ankle.[18] Experiments performed at the National Hansen's Disease Center in Carville, Louisiana, linked repetitive mild to moderate stress on the insensitive limb with a generalized increase in local skin temperature.[6] Unfortunately, the sensory neuropathy often renders the person unaware of the bony destruction that takes place during ambulation. Signs of an acute Charcot's joint may be absent. Due to the lack of subjective symptoms, it may also be difficult to objectively grade the course of healing of acute Charcot arthropathy.

Infrared dermal thermography provides an inexpensive means of measuring local skin temperatures to evaluate inflammation in the Charcot foot. Several investigators have reported on the use of foot skin temperature monitoring as a potentially useful tool in diabetic patients in the detection of neuropathic lower extremity sequelae. Chan and colleagues demonstrated that patients with diabetes and painful neuropathy had higher forefoot skin temperatures than control subjects without diabetes.[11] Benbow et al. in 1994 found that thermography can be used to successfully predict neuropathic plantar foot ulceration.[7] They measured the mean plantar foot temperature in 50 patients with painful diabetic sensory-motor neuropathy, and found that the patients who developed a foot ulcer had significantly higher mean plantar foot temperatures.

Armstrong et al. in *Physical Therapy* (1997) compared skin temperatures in patients with asymptomatic peripheral sensory neuropathy, patients with new-onset neuropathic ulcers, and patients with Charcot's arthropathy, using the contralateral limb as a control.[5] Thus, each patient served as his or her own physiological control. They evaluated 143 subjects (96 male, 47 female) with type II diabetes mellitus who had an average age of 63.9 years, at a multidisciplinary tertiary care diabetic foot center between 1993 and 1995. They used an Exergen DT 1001 infrared skin temperature probe. This device measures approximately 1.0 cm squared area of skin and is held approximately 0.5 cm from the skin surface. It is accurate to within +/- 0.2 °F. Readings were recorded from six sites on the soles of both feet: the hallux, the first metatarsal-cuneiform joint, the cuboid, and the first, third and fifth metatarsal heads. The authors found significant differences in skin temperature between the affected foot and the contralateral non-affected foot among patients with neuropathic ulcers (5.6 °F) with skin temperatures

> Infrared dermal thermography provides an inexpensive means of measuring local skin temperatures to evaluate inflammation in the Charcot foot.

highest at the ulcer site in 95%. They also found that, of the 44 patients with neuropathic ulcers that had healed, 5 (11.4%) experienced re-ulceration at the site of the previous ulcer. They noted that skin temperatures taken during the visit prior to re-ulceration were higher on the pre-ulcerative limb than on the contralateral limb (89.6 +/- 1.2 °F versus 82.5 +/- 2.9 °F). The authors felt that elevated skin temperatures may be predictive of future ulceration. In patients with Charcot's arthropathy, they found a difference of 8.3 °F between the affected foot and the nonaffected foot. The authors found that the site of maximum skin temperature correlated with the anatomic site of maximum bony involvement (radiographically) in all patients. In patients with diabetic peripheral sensory neuropathy and no additional pathology, the authors found no difference in temperature between the feet.

Armstrong and Lavery in the *Journal of Rehabilitation Research and Development* (July 1997) used infrared thermography to monitor the clinical course of 39 diabetic patients with acute Charcot's arthropathy.[4] The authors monitored foot skin temperatures using a portable infrared thermometric probe. They used the contralateral limb as a physiological control and compared skin temperatures on the affected and control foot. A positive skin temperature gradient implied that skin temperatures were greater on the foot with acute Charcot's arthropathy than the contralateral foot. Subjects were treated with a protocol involving serial total contact casting with progression to removable cast walkers and finally to prescription therapeutic shoe gear. Casting was discontinued based on clinical, radiographic, and thermometric (temperature equilibration with the contralateral limb) signs of quiescence. The authors found that the site of maximum skin temperature gradient correlated to the site of maximum Charcot arthropathy (radiographically) in 92% of cases. The skin temperatue gradient gradually decreased during total contact casting therapy. Following transition to prescription shoe gear, the skin temperature gradient was near zero. The subjects that re-ulcerated during the follow-up period all showed significantly elevated skin temperature gradients on the visit prior to ulceration. The authors noted that there were no recurrences of Charcot's fractures in an average follow-up period of two years. They felt that this was due to their aggressive early intervention when subjects demonstrated an increase of more than 4 °F in the previously affected foot compared to the contralateral foot. These patients were immobilized in a removable cast walker or total contact cast and instructed to sharply limit their activity.

As the clinical changes that precede ulceration may be subtle and difficult to appreciate in persons with diabetes, the authors feel that infrared dermal thermography can be useful to detect early skin temperature changes that may be nonpalpable and that may be predictive of a new ulceration. Thus, early intervention such as the use of therapeutic footwear or total contact casting can be employed. In addition, as 20% to 58% of patients with healed neuropathic ulcers redevelop another ulcer within one year, thermography can be helpful in monitoring skin temperatures during subsequent follow-up visits. An increase in local skin temperature after wound healing raises the suspicion of a re-ulceration and the need for more aggressive measures. Thermography could also be useful in monitoring the progression of Charcot's fractures through the major clinical phases. In the post-acute phase, when gross signs of inflammation have resolved, skin temperature measurements can be combined with clinical and radiographic signs of resolution, to gauge when guarded weight bearing can begin. This would then prevent premature return of the patient to an activity that could trigger damage to the foot.

> Daily self-monitoring with a handheld infrared skin thermometer can significantly reduce the number of diabetic foot complications.

Lavery et al. in *Diabetes Care* (2004) have also shown that home monitoring of foot skin temperatures can help prevent ulceration.[27] In individuals at high risk for lower-extremity ulceration and amputation, they found that daily self-monitoring with a handheld infrared skin thermometer can significantly reduce the number of diabetic foot complications. The patients were instructed to measure the skin temperature of their soles every morning and night. If one foot was >4 °F warmer than the other, it was considered to be "at risk" of ulceration due to inflammation. If their foot temperatures were elevated, the patients were instructed to reduce their activity and contact the study nurse.

Sun et al. in *Foot and Ankle International* (2005) addressed the lack of formal, inter-institutional thermal measurement procedures and proposed a standardized protocol to quantify foot temperature.[39] They found that the temperature in each of six plantar subregions varied as a function of time, and that the mean temperature of the entire plantar area was more stable than the individual subregions. Thus, they recommended that the mean plantar temperature be used as a more reliable indicator of thermoregulatory function. The researchers were also able to confirm that the variation of temperature with time stabilizes after 15 minutes. While it had long been recommended that subjects rest through a 15 minute equilibration period prior to infrared imaging, Sun et al. established that in the foot, the mean plantar temperature reaches equilibrium after 15 minutes.

In 2006, Bharara et al. in *Lower Extremity Wounds* argued in support of the whole-field imaging capabilities of thermography, and its potential for identifying pre-ulcerous changes in the diabetic patient at increased risk of foot ulceration.[8]

Ohsawa et al. in *Archives of Orthopedic Trauma Surgery* (2001) amputated 35 limbs of 27 patients with diabetic feet from March 1988 to March 1998.[30] The amputation level of the limb was, in part, determined by skin thermography. Logistic regression analysis revealed that significant risk factors for re-amputation included lower temperature at the amputation site, being female, or being elderly. The authors concluded that skin thermography was one of the effective determinants to guide amputation level in order to limit the need for re-amputation.

Hyperhidrosis

Hyperhidrosis is a disorder of the autonomic nervous system resulting in excessive sweating. It can be both socially and occupationally distressing. Americans alone spend hundreds of millions of dollars annually in sweat-reducing products. As hyperhidrosis is in the spectrum of autonomic mediated disorders, studies have evaluated the role of infrared imaging in diagnosing hyperhidrosis as well as in monitoring the effects of a surgical sympathetic block on sudomotor and vasoconstrictor function.[26,32,33]

Sweating is a physiological function of the sympathetic nervous system that involves the active secretion of a watery fluid onto the body surface from either eccrine or apocrine sweat glands. Sweating is under the control of both sympathetic innervation and circulating catecholamines. In the sweating process, afferent impulses from sensors for skin and core temperature are initially received by the hypothalamus. Then, efferent impulses from the preoptic area of the anterior hypothalamus travel along sympathetic fibers to innervate eccrine and apocrine sweat glands.[28] The sympathetic nerves originate in the spinal cord between the levels of T-1 and L-2, and the distribution of sympathetic outflow is segmental. The T-2 spinal segment is thought to be the key source of preganglionic outflow to the sweat glands of the upper extremity.

Surgical sympathectomy has been the treatment of choice for moderate to severe palmar and facial hyperhidrosis, as well as axillary hyperhidrosis if combined with palmar sweating. The key in surgical therapy of palmar hyperhidrosis involves the complete denervation of the sympathetic nervous system in the upper extremity. This is achieved by surgically dividing the sympathetic chain above the T-2 ganglion and below the T-3 ganglion and excising the T-2 and T-3 ganglia along with the corresponding spinal nerves.[31] Overall patient satisfaction with surgical management of palmar hyperhidrosis is high. In the immediate postoperative period, 92% to 98% of patients have complete relief of symptoms. However, in 28% to 53% of these patients, some palmar sweating recurs after one year, although it is rarely disabling. Complications of sympathectomy are uncommon and include infection, pneumonia, Horner's syndrome, and compensatory hyperhidrosis in nondenervated areas. Endoscopic thoracic sympathectomy is widely used instead of conventional thoracic sympathectomy, resulting in less trauma and fewer complications.[38]

Rosenblum et al. in *Angiology* (1994) used infrared thermography to evaluate a 39-year-old male with severe plantar hyperhidrosis.[32] Schick et al. in *Neurology* (2003) used infrared thermography to assess the effects of a surgical block of the sympathetic chain at the T2 level on 61 patients with palmar hyperhidrosis.[33] They studied the re-warming kinetics following immersion of both hands in ice water at 4 °C for 30 seconds; preoperatively, 2 days postoperatively and 3 months postoperatively. Results were compared to data from 28 healthy controls. The authors found that prior to receiving the sympathetic block, the re-warming kinetics were significantly slower in the patients than in the controls. Two days following the sympathetic block, baseline skin temperatures in the patients had increased by 5 °C and the re-warming was massively accelerated. Three months post-sympathetic block, re-warming kinetics were still accelerated in 36 hands, unchanged from the preoperative condition in 42 hands, and had worsened in 12 hands. In all but one patient, palmar sweating was massively reduced, irrespective of the re-warming kinetics. The authors concluded that T2 sympathectomy leads to a long-lasting inhibition of palmar sweating, which does not correlate to the loss of vasoconstriction.

Kruger et al. in *Clinical Autonomic Research* (2003), in a prospective study, compared the effects of an endoscopic sympathetic block (ESB) at the level of the second (T2) and fourth (T4) thoracic ganglion on vasoconstriction and palmar sweating.[26] Computer assisted infrared thermography was used to measure sympathetic vasoconstriction in 22 hyperhidrosis patients who had undergone ESB at T2 or T4. Palmar sweating was assessed by sudometry. Ice water immersion of the hands was performed before, two days, and three months post-sympathetic block. The authors found that following ESB, rewarming was accelerated in both the T2 and T4 patients, but was significantly faster in the T2 group. At three months postoperatively, rewarming was still significantly faster in the T2 group. These effects were more pronounced in the fingertips than the dorsum of the hands. Sudomotor function was blocked in all T2 patients, but two T4 patients did not show an effect on sudomotor function postoperatively. The authors felt that the normalization of rewarming kinetics was explained by remaining fibers, denervation hypersensitivity, stimulation of catecholamine receptors, or neuronal reorganization. They felt that the effect of ESB of T4 on sudomotor function has yet to be proven.

Severe Acute Respiratory Syndrome

There has been a great deal of interest in the application of infrared technology as a screening tool for severe acute respiratory syndrome (SARS). Fever greater than 38 ºC is a cardinal sign for severe acute respiratory syndrome.[14] Remote sensing infrared thermography (IRT) has been advocated as a means for screening for fever in travelers at airports and border crossings.

Chan et al. in *Journal of Travel Medicine* (2004) evaluated the feasibility of IRT imaging to identify subjects with fever and to establish the optimal instrumental configuration and validity for such testing.[12] They evaluated 176 subjects and obtained IRT readings from various parts of the front and sides of the face, at distances of 0.5m and 1.5m. They compared their results with concurrently obtained body temperatures using conventional means (aural tympanic IRT and oral mercury thermometry). The authors found that ear IRT readings at 0.5m yielded the narrowest confidence intervals and could be used to predict conventional body temperature readings at < or = 38 ºC with a sensitivity of 83% and a specificity of 88%. They felt that their findings could have important implications for walk-through IRT scanning/screening systems at airports and border crossings.

> Remote sensing infrared thermography (IRT) has been advocated as a means for screening for fever in travelers at airports and border crossings.

Chiu et al. in *Asia-Pacific Journal of Public Health* (2005) used digital infrared thermal imaging to conduct mass screenings of patients and visitors who entered Taipei Medical University in Taipei, Taiwan from April 13 to May 12, 2003.[14] A total of 72,327 outpatients and visitors were screened, of which 305 (.42%) were detected as being febrile. Of these, three probable SARS patients were identified. The authors' findings suggested that infrared thermography is an effective and reliable tool suitable for mass screening for fever in the initial phase of screening for SARS.

References

1. al-Awami M, Schillinger M, Maco T, Pollanz S, and Minor E. Low level laser therapy for the treatment of primary and secondary Raynaud's phenomenon. *VASA,* 2004;33(1):25-9.

2. Albert SM, Glickman M, and Kallish M. Thermography in Orthopedics. *Ann NY Acad Sci,* 1964;121:157-70.

3. Anderson ME, Moore TL, Lunt M, Herrick AL. The 'distal-dorsal difference': a thermographic parameter to which to differentiate between primary and secondary Raynaud's phenomenon. *Rheumatology* (Oxford), 2007;46(3):533-8.

4. Armstrong DG and Lavery LA. Monitoring healing of acute Charcot's arthropathy with infrared dermal thermometry. *J Rehab Res Dev,* 1997;34:317-21.

5. Armstrong DG, Lavery LA, Liswood PJ, Todd WF, and Tredwell JA. Infrared Dermal Thermometry for the high-risk diabetic foot. *Phys Ther,* 1997;77:169-75.

6. Beach RB and Thompson DE. Selected soft tissue research: an overview from Carville. *Phys Ther,* 1979;59:30-5.

7. Benbow SJ, Chan AW, Bowsher DR, Williams G, and Macfarlane IA. The prediction of diabetic neuropathic plantar foot ulceration by liquid crystal contact thermography. *Diabetes Care,* 1994;17:835-9.

8. Bharara M, Cobb JE, and Claremont DJ. Thermography and thermometry in the assessment of diabetic neuropathic foot: A case for furthering the role of thermal techniques. *Int J Low Extrem Wounds,* 2006;5:250-60.

9. Brelsford KL and Uematsu S. Thermographic presentation of cutaneous sensory and vasomotor activity in the injured peripheral nerve. *J Neurosurg,* 1985;62:711-5.

10. Brown RKJ, Bassett LW, Wexler CE, and Gold RH. Thermography as a screening modality for nerve fiber irritation in patients with low back pain. *Modern Medicine* (Special Suppl), 1987;55:86-8.

11. Chan AW, MacFarlane IA, and Bowsher DR. Contact thermography of painful diabetic neuropathic foot. *Diabetes Care,* 1991;14:918-22.

12. Chan LS, Cheung GY, Lauder D, Kumana CR, and Lauder D. Screening for fever by remote-sensing thermographic camera. J *Travel Med,* 2004;11(5):273-9.

13. Cherkas LF, Carter L, Spector TD, Howell KJ, Black CM, and MacGregor AJ. Use of thermographic criteria to identify Raynaud's phenomenon in a population setting. *J Rheumatol,* 2003;30(4):720-2.

14. Chiu WT, Lin PW, Chiou HY, Lee WS, Lee CN, Yang YY, Lee HM, Hsieh MS, Hu CT, Deng WP, and Hsu CY. Infrared thermography to mass-screen suspected SARS patients with fevers. *Asia Pac J Public Health,* 2005;17(1):26-8.

15. Clark S, Hollis S, Campbell F, Moore T, Jayson M, Herrick A. The "Distal-Dorsal Difference" as a possible predictor of secondary Raynaud's phenomenon. *J Rheumatol,* 1999;26:1125-8.

16. Darton K and Black CM. Pyoelectric Vidicon Thermography and cold challenge quantify the severity of Raynaud's phenomenon. *Br J Rheumatol,* 1991;30:190-5.

17. Dieppe PA, Sathapatayavongs B, Jones HE, Bacon PA, and Ring EF. Intra-articular steroids in osteoarthritis. *Rheumatol Rehabil,* 1980;19:212-7.

18. Edmonds ME. The neuropathic foot in diabetes. Part 1: Blood flow. *Diabet Med,* 1986;3:111-5.

19. Fischer AA and Chang CH. Deep tissue temperature and thermography in trigger points and painful areas. Third International Congress of Thermography. Bath, England, March 29-April 2, 1982.

20. Fischer, AA. "Diagnosis and management of chronic pain in physical medicine and rehabilitation," in *Current Therapy in Physiatry,* A.P. Ruskin, Ed. WB Saunders, Philadelphia, 1984;123-45.

21. Gold JE, Cherniack M, and Buchholz B. Infrared thermography for examination of skin temperature in the dorsal hand of office workers. *Eur J Appl Physiol,* 2004;93(1-2):245-51.

22. Hadler NM. Arm pain in the work place. A small area analysis. *J Occup Med,* 1992;34(2):113-9.

23. Hakguder A, Birtane M, Gurcan S, Kokino S, and Turan FS. Efficacy of low level laser therapy in Myofascial Pain Syndrome:An Algometric and Thermographic Evaluation. *Lasers Surg Med,* 2003;33:339-43.

24. Herrick RT and Herrick SK. Thermography in the detection of carpal tunnel syndrome and other compressive neuropathies. *J Hand Surg,* 1987;12A(2 Pt 2): 943-9.

25. Hoffman RM, Kent DL, and Deyo RA. Diagnostic accuracy and clinical utility of Thermography for lumbar radiculopathy—A Meta-Analysis. *Spine,* 1991;16:623-8.

26. Kruger S, Fronek KS, Schmelz M, Horbach T, Hohenberger W, and Schick CH. Differential effects of surgical sympathetic block at the T2 and T4 level on vasoconstrictor function. *Clin Auton Res,* 2003;13(1):179-83.

27. Lavery LA, Higgins KR, Lanctot DR, Constantinides GP, Zamorano RG, Armstrong DG, Athanasiou KA, and Agrawal CM. Home monitoring of foot skin temperatures to prevent ulceration. *Diabetes Care,* 2004;27: 2642-7.

28. Manusov EG and Nadeav MT: Hyperhidrosis: A management dilemma. *J Family Pract,* 1989;28:412-5.

29. Ming Z, Zaproudina N, Siivola J, Nousianen U, and Pietikainen S. Sympathetic pathology evidenced by hand thermal anomalies in carpal tunnel syndrome. *Pathophysiology,* 2005;12(2):137-41.

30. Ohsawa S, Inamori Y, Fukuda K, and Hirotuji M. Lower limb amputation for diabetic foot. *Arch Orthop Trauma Surg,* 2001;121(4):186-90.

31. Riolo J, Gumucio CA, Young AE, and Young VL. Surgical management of palmar hyperhidrosis. *South Med J,* 1990;83:1138-43.

32. Rosenblum JA, Cohen JM, and Lee MHM. Hyperhidrosis — a case history. *Angiology,* 1994;45(1):61-4.

33. Schick CH, Fronek K, Held A, Birklein F, Hohenberger W, and Schmelz M. Differential effects of surgical sympathetic block on sudomotor and vasoconstrictor function. *Neurology,* 2003;60(11):1770-6.

34. Schufried O, Vacariu G, Lang T, Korpan M, Kiener HP, and Fialka-Mosen V. Thermographic parameters in the diagnosis of secondary Raynaud's phenomenon. *Arch Phys Med Rehabil,* 2000;81(4):495-9.

35. Sharma SD, Smith EM, Hazleman JR, and Jenner JR. Thermographic changes in keyboard operators with chronic forearm pain. *BMJ,* 1997;314:118.

36. So YT, Aminoff MJ, and Olney RK. The role of thermography in the evaluation of lumbosacral radiculopathy. *Neurology,* 1989;39:1154-8.

37. So YT, Olney RK, and Aminoff MJ. Evaluation of thermography in the diagnosis of selected entrapment neuropathies. *Neurology,* 1989;39:1-5.

38. Stolman LP. Treatment of Hyperhidrosis. *Current Therapy,* 1998;16(4):863-7.

39. Sun PC, Jao SHE, and Cheng CK. Assessing foot temperature using infrared thermography. *Foot and Ankle Int,* 2005;26:847-53.

40. Uematsu S. Thermographic imaging of cutaneous sensory segments in patients with peripheral nerve injury. *J Neurosurg,* 1985;62:716-20.

41. Varju G, Pieper CF, Renner JB, and Kraus VB. Assessment of hand osteoarthritis: correlation between thermographic and radiographic methods. *Rheumatology,* 2004;43:915-9.

42. Von Bierbrauer A, Schilk I, Lucke C, and Schmidt JA. Infrared thermography in the diagnosis of Raynaud's phenomenon in vibration-induced white finger. *VASA,* 1998;27:94-9.

43. Weinstein G. The diagnosis of trigger points by thermography. Academy of Neuro-Muscular Thermography Clinical Proceedings, Postgraduate Medicine Custom Communications, (March) 1986.

44. Weinstein SA. A comparison of thermography with EMG, CAT scanning, myelography and surgery in 250 patients with low back symptoms. 13th Annual Meeting, Academy of Thermology, Washington DC, (June) 1984.

Thermography in the Evaluation of Decubitus Ulcers

ADEEL AHMAD, MD

A pressure ulcer is a wound that arises from excess pressure or friction damaging skin and underlying tissues. A pressure ulcer is also commonly referred to as a decubitus ulcer or bedsore.[5] As health care costs rise, average life expectancies increase, and populations in long-term care facilities continue to grow, increased attention to this dilemma is necessary. Current treatment of pressure sores is in excess of $1 billion annually.[12] The fact that no pharmacological intervention has yielded conclusive evidence for effectiveness against pressure ulcers emphasizes the need for prevention and early identification.[5]

This chapter will seek to address the use of thermography (computerized infrared imaging or CII) in assessing pressure ulcers as well as its use in its prevention. Clinical applications discussed in the current literature will be discussed.

Background

Roughly 10% of acute care patients in the United States acquire pressure ulcers during their hospital stays.[2] Treatment can be difficult and costly. Pressures placed onto skin, muscle, and soft tissues can exceed the average capillary pressure of 32 mm Hg. This pressure can occur to varying sites with patients in a lying or seated position. Those patients with normal sensation, mobility, and mental capacity are able to alter and shift positions so as to avoid pressure buildup that would otherwise result in irreversible tissue damage.[12] Individuals who are elderly, neurologically impaired, or acutely hospitalized are at increased risk of

> Roughly 10% of acute care patients in the United States acquire pressure ulcers during their hospital stays.

developing pressure ulcers. Pressure ulcers tend to occur over the lower body. In decreasing frequency, the most common sites of ulcers have been stated as occurring in the sacral region, the heels, elbows, ankle, trochanters, ischia, knees, scapulas, shoulders and occiput.[5]

Risk Factors

Cannon and Cannon[5] have described intrinsic and extrinsic risk factors that predispose to the development of pressure ulcers. Among the intrinsic factors is age, likely secondary to age-related changes of the skin. Conditions such as diabetes and spinal cord injury, which may impair sensation or pain perception, place individuals at increased risk. Other intrinsic factors include lack of consciousness or diminished awareness. Poor nutritional status and vascular disease, which may both impair wound healing, have also been implicated as affecting pressure wound development.[5]

Extrinsic risk factors for developing pressure ulcers are those that result from the environment around the wound. The most important factor is pressure exceeding the normal capillary pressure of 32 mm Hg. Revis states that irreversible changes may occur in as little as two hours of uninterrupted pressure.[12] Friction and shearing forces also contribute to disruption of normal circulation leading to tissue ischemia.[9]

Etiology

The exact etiology of pressures is unclear. However, it is felt that the causes are multifactorial and likely due to a combination of pressure and shear forces at the level of the skin and the underlying architecture. The underlying tissues are more likely to be affected by a disruption of blood flow than the epidermis.[1] This implies that damage may be more severe than that revealed on a gross visual exam of the skin.

Staging

The National Pressure Ulcer Advisory Panel (NPUAP) staging system from the 1989 Consensus Development Conference describes four categories of increasingly severe pressure ulcers.[10]

- Stage 1 pressure ulcers have observable pressure related changes to intact skin and may include changes in skin temperature, tissue consistency (firm or boggy feel), and sensation (pain, pruritic). Stage 1 ulcers have defined areas of persistent redness in lightly pigmented skin and persistent red, blue, or purple hues in darker skin tones.
- Stage 2 ulcers involve partial thickness skin loss of the epidermis, dermis or both. The ulcer is superficial and presents as an abrasion, blister, or shallow crater.
- Stage 3 ulcers are full thickness skin wounds that involves damage or necrosis of subcutaneous tissue without involving the underlying fascia. These wounds present as deep craters that may or may not undermine adjacent tissue.
- The last category, stage 4 ulcers, describes wounds with full thickness skin loss with extensive destruction, tissue necrosis, damage to muscle, bone, or supporting structures. Sinus tracts may also be associated with stage 4 ulcers.[10]

Treatment

A review of the current literature by Cannon and Cannon shows that, currently, no pharmacologic intervention has been conclusively effective for pressure ulcers.[5] Treatment begins with removal of the source of pressure, assessment of intrinsic and extrinsic risk factors, and staging of ulcers. Attention to improving nutrition, specifically increasing protein content, has demonstrated benefit. Wound care involving debridement and specialized wound dressings represents the next step in treatment. These dressings exist in a wide variety and cost, and are made for ulcers of all stages and drainage characteristics. Growth factors have been studied with only platelet-derived growth factor demonstrating some benefit. Its use remains controversial. As of now, no topical growth factors carry FDA-approved labeling for use in pressure ulcer treatment.[5]

No pharmacologic intervention has been conclusively effective for pressure ulcers.

Reactive hyperemia and thermography

Reactive hyperemia is a reflex action in capillary beds of skin that occurs after blood flow has been restored following a period of occlusion.[8] A strong reactive hyperemic response generally denotes healthy circulatory function. In these tissues, the capillary shunts in the dermis continuously open and close in a cyclic fashion, thus regulating blood flow to the region. If tissues become ischemic, the capillary remain open for a larger percentage of time. If the ischemia is secondary to prolonged mechanical compression, the relief of pressure will result in maximal blood flow returning to the affected areas, producing marked erythema.[8] This physiologic response to pressure can be measured indirectly with thermography.[7]

Hansen et al. examined the use of cutaneous reactive hyperemia to assess wound severity in newly formed temperature-modulated pressure ulcers.[8] They sought to determine whether the type of hyperemic response could guide appropriate care. Pressures were induced in animal models, responses were measured by color and thermographically, and finally these measurements were compared with histologic specimens of the wounds themselves. The authors found that differentiating mild and deep wounds on the basis of color to be difficult. Deep wounds showed severe histologic damage with little involvement at the skin surface where color was measured. The thermographic approach was able to establish a clear wound category. The mild injuries tended to be warmer than reference while the moderate damage injuries corresponded with near normal skin temperature. The severely injured sites, again as established on biopsy, appeared cooler, likely secondary to severe damage to the vasculature in the region. This damage prevented an influx of blood flow which is characteristic of a normal hyperemic response.[8]

Tools to measure reactive hyperemia

As stated above, reactive hyperemia is observed as the physiologic response to pressure. It can be measured both indirectly and directly. Laser Doppler flowmetry provides a direct measure by assessing movement of blood cells within skin micro vessels. Thermography, on the other hand, measures heat brought to the skin by blood flow from the body's core.[4]

Thermography's advantage over laser Doppler flowmetry lies in its indirect nature. Motion artifacts, skin-contacting probes, and skin pigmentation potentially confound laser Doppler flowmetry's data. Thermographic assessment is free of these problems. Still, a temperature-controlled environment and uniform airflow are essential to accurate thermographic data collection.[4]

The use of thermography to assess severity of pressure ulcers

Hansen et al. used thermography to evaluate 1) wounds at thermal equilibrium with normal temperature surroundings and 2) wound response to focal cooling.[7] The authors developed a series of experiments using a porcine model in which standardized wounds were created. These wounds were then evaluated thermographically and histologically. Through correlations of histological classification of wound status (severity and depth) and the thermographic data, the authors determined that deep-tissue injuries were easily distinguishable from shallow injuries by their response to focal cooling.[7]

The first experimental series yielded pronounced differences in skin surface temperature depending on the wound depth and time following infliction of injuries. Surface-only wounds had an initial temperature that was much lower than that of the other wound types. However, these wounds recovered to near normal temperatures. Full-thickness injuries showed a steady decrease in temperature. The deep-only injuries, which did not show any visual damage at the skin surface, did not show any reproducible changes in surface temperature. Many of the wounds displayed what the authors described as "periwound temperature elevation" (PWTE). These wounds consisted of a cool wound center surrounded by a warmer wound margin. The PWTE was strongly associated with full-thickness injuries and was occasionally associated with shallow wounds. The wounds exhibiting PWTE were more severe than non-PWTE injuries.[7]

Hansen et al.'s second series revealed that a wound's surface temperature varies in response to focal cooling. The wounds were focally cooled for a fixed amount of time and then allowed to obtain thermal equilibrium. The mean temperature travel (calculated from the sum of the mean temperature drop and rise) showed statistically significant differentiation in wound types from days 4-9 following wound infliction.[7]

The authors concluded that thermographic temperature measurements following wound cooling may allow detection of deep tissue

> The authors concluded that thermographic temperature measurements following wound cooling may allow detection of deep tissue injuries that are "hidden" and not visually detectable.

injuries that are "hidden" and not visually detectable and that PWTE may indicate severe injury and correlates strongly with full-thickness wounds.[7]

Use of thermography in pressure ulcer clinical research

Newman and Davis evaluated thermography as a predictor of sacral pressure sores.[11] Ninety-one patients were studied. The patients had been admitted to a nursing home with no visual evidence of damaged sacral skin. Nineteen percent of these patients showed an abnormal thermal pattern consistent with occult skin damage. Thirty-five percent of these patients went on to develop an apparent pressure sore within 10 days of admission.[11]

Barnett and Ablarde compared the utility of thermography to determine the ideal patient positioning on a mattress.[3] The authors compared four different side-lying positions and measured the pressure at the bony prominence (greater trochanter). The pressure measurements indicated that the greatest pressures were experienced with both legs extended and the least when the lower leg was extended and the upper leg flexed 45 to 90 degrees. This correlated with thermographic studies showing greatest redness and reactive hyperemia over the subjects' trochanters and iliac crests when the bilateral hips were extended. Based on this data, the authors further suggested that turning patients every two hours may not be sufficient and that positioning with the legs in the extended position should be minimized.[3]

Ferrain and Ludwig used thermography to assess the distribution of temperature over the surface of wheelchair cushions.[6] The authors viewed the thermal transient of the body (the heating phase) and after its use (the cooling phase). Air-filled cushions demonstrated the fastest thermal transient while gel-cushions showed the slowest transient. The surfaces of the cushions were also studied. Foam surfaces were found to have the highest peak temperatures, followed by air-filled cells and bubble-shaped surfaces. Also significant was the finding that temperature under the thighs was greater than under the ischial areas with all cushions.[6]

Conclusions

Pressure ulcers provide a serious concern for many patients. Identification of intrinsic and extrinsic risk factors and preventative care is essential. Thermography's utility in evaluating pressure ulcers lies in its ability to indirectly measure reactive hyperemia. Clinically, thermography can help determine patients at risk of developing pressure ulcers and can evaluate wound severity and depth. Responses to focal cooling and PWTE may provide useful clinical data in selected patients. Potential research applications using this tool include evaluating patient-surface interactions and have thus far yielded information on topics such as patient positioning and wheelchair cushions.

> Clinically, thermography can help determine patients at risk of developing pressure ulcers and can evaluate wound severity and depth.

References

1. Allman RM. Pressure ulcers among the elderly. *N Engl J Med*, 1989;320:850-3.
2. Barczak CA, Barnett RI, Childs EJ, and Bosley LM. Fourth National Pressure Ulcer Prevalence Survey. *Adv Wound Care*, 1997;10(4):18-28.
3. Barnett RI and Ablarde JA. Skin Vascular Reaction to Standard Patient Positioning on a Hospital Mattress. *Adv Wound Care*, 1994;7(1):58-65.
4. Barnett RI and Shelton FE. Measurement of Support Surface Efficacy: Pressure. *Adv Wound Care*, 1997;10(7):21-9.
5. Cannon CC and Cannon JP. Management of pressure ulcers. *Am J Health-Syst Pharm*, 2004;61:1895-907.
6. Ferrain M and Ludwig N. Analysis of thermal properties of wheelchair cushions with thermography. *Med Biol Eng Comput*, 2000;38:31-34.
7. Hansen GL, Sparrow EM, Kaleita AL, and Iaizzo PA. Using Infrared Imaging to Assess the Severity of Pressure Ulcers. *Wounds*, 1998;10(2):43-53.
8. Hansen GL, Sparrow EM, Kommamuri N, and Iaizzo PA. Assessing wound severity with color and infrared imaging of reactive hyperemia. *Wound Rep Reg*, 1996;4:386-392.
9. Meijer JH, Germs PH, Schneider H, and Ribbe MW. Susceptibility to Decubitus Ulcer Formation. *Arch Phys Med Rehabil*, 1994;75:318-323.

10. National Pressure Ulcer Advisory Panel. Staging Report. 2003. http://www.npuap.org/positn6.html.

11. Newman P and Davis NH. Thermography as a predictor of sacral pressure sores. *Age and Ageing,* 1981;10:14-18.

12. Revis DR. Decubitus Ulcers. *Emedicine,* 2005. http://www.emedicine.com/med/topic2709.htm.

Thermography and Orofacial Pain

Terri A. Norden, DDS, MD
Jeffrey M. Cohen, MD

Pain in the facial region can be among the most difficult areas for clinicians to evaluate as the causes may be dental or medical in origin and very elusive. Misdiagnosis is common, and often pain continues after treatment. Patients can undergo occlusal adjustments and tooth extractions needlessly. Thermography as a tool in evaluating dental symptoms and aiding in diagnosis was first described by Crandell in 1966.[3]

When performing facial thermography, it is essential to minimize external influences on skin temperature. Facial thermography protocols call for stabilization of the surrounding environment prior to initiating the study. Basic guidelines for conducting facial thermographic examinations were published by the American Academy of Neuro-muscular Thermography in 1986. Their guidelines include:[17]

- a draft-free environment, with no windows and closed doors
- a temperature controlled room with controlled relative humidity
- pulling the patient's hair back from the face
- verification that the patient's face is clear of make-up by wiping it with a damp cloth
- defer patients with sunburn for at least 10 days
- a small hand-held electric fan for cooling of the face (25s) prior to imaging
- a 15-minute equilibration period to allow the patient to reach a stable physiologic state

A facial thermographic examination is generally tolerated by patients, noninvasive, painless, and can be completed in less than 30 minutes.

The Normal Facial Thermogram

Normal values of heat emission from the face in asymptomatic individuals have been established by analyzing patterns of temperature distribution on facial thermograms.

Initial studies by Gratt and Sickles in 1995[5] identified a high degree of thermal symmetry about the face. Twenty-five anatomic regions including the orbits, upper lip, lower lip, chin and cheek were studied in 102 subjects. Their data revealed that the mean thermal differences (area ΔT) of each of the 25 regions, right-versus-left-side, in individual subjects were within 0.1 °C of each other. In addition, their study identified significant differences in the absolute facial temperatures between men and women, with men having higher overall facial temperatures in the 25 anatomic zones.

The Abnormal Facial Thermograph

In 1983, Soffin et al. used thermography to evaluate oral inflammatory conditions.[14] They found that the thermogram was helpful in delineating the involved side from the uninvolved side in a majority of the cases they studied. Gratt et al. in 1996 performed a prospective, matched study consisting of 164 dental patients and 164 matched (control) subjects.[7] Building upon their previous study of normal subjects,[5] they grouped subjects by area ΔT values (defined as the mean right-versus-left side thermal difference in a given area). The authors found that patients with "hot" thermograms (selected area ΔT > +0.35 °C) had conditions such as maxillary sinusitis, sympathetically maintained pain, peripheral nerve mediated pain, temporomandibular joint (TMJ) osteoarthropathy, and TMJ with acute internal derangement. Patients with "cold" thermograms (selected area ΔT < -0.35 °C) were found to have peripheral nerve mediated pain and sympathetically independent pain. Subjects classified as having "normal" thermograms (selected area ΔT between 0.0 and ±0.25 °C) were found to have the clinical diagnoses of cracked tooth syndrome, trigeminal neuralgia, pre-trigeminal neuralgia or psychogenic pain. (Figure 1)

There was a 92% agreement (301 of 328) in classifying orofacial pain patients versus their matched controls using this system of thermal classification. As a result of this study, "area ΔT" became an important diagnostic parameter in facial thermography.

Classification	ΔT	Diagnosis
Normal	±0.25 °C	• cracked tooth syndrome • trigeminal neuralgia • pretrigeminal neuralgia • psychogenic pain
Hot	> +0.35 °C	• sympathetically maintained pain • peripheral nerve mediated pain • TMJ osteoarthropathy • maxillary sinusitis
Cold	< -0.35 °C	• peripheral nerve mediated pain • sympathetically independent pain
Equivocal	±(0.26 – 0.35 °C)	

Figure 1. *Classification System of ΔT Measurements* [7]

Temporomandibular Joint Disorders

The temporomandibular joint (TMJ) is a very frequent cause of facial pain. Internal derangement occurs when pathology affects the normal relationship between the mandibular condyle, the glenoid fossa, the articular disc, and the articular eminence. The joint may be associated with otitis media or with cervical pathology.

In Gratt and Sickles' 1993[6] study of asymptomatic TMJ patients, the authors found that facial thermograms were symmetric over the bilateral TMJs, with mean temperature differences of less than 0.2 °C between the right and left side. A 1994 study by Gratt et al. of 30 patients with internal derangement of the TMJ, however, revealed low levels of thermal symmetry. The authors found temperature differences (area ΔT values) of 0.4 °C to 0.8 °C when comparing the symptomatic to asymptomatic side.[8]

Pogrel et al.[12] used thermography to study the relationship between cervical region dysfunction and temporomandibular dysfunction (TMD). They evaluated 22 patients with unilateral (TMD) and 22 normal controls and found a statistically significant increase in temperature asymmetry in the upper back and neck of TMD patients (0.78 °C) versus control subjects (0.13 °C), regardless of the presence of active upper back and neck pain. They found a 95.5% correlation between the presence of TMD and an increase in ipsilateral trapezius muscle temperature.

A similar study by Pogrel et al.[11] used thermography to help distinguish between problems that were primarily due to masseter dysfunction associated with myofascial pain and problems due to internal derangment of the TMJ. Patients with painful internal derangements of the TMJ were found to have warm areas over the joint, while those with myogenic facial pain symptoms had variable hot and cold spots over the masseter muscle. Following nonsurgical treatment of these conditions, the thermograms of both returned to normal.

In 1995, Canavan and Gratt used thermography to differentiate between asymptomatic control subjects and patients with mild to moderate TMJ disorders.[2] Twenty-four asymptomatic controls and twenty patients with TMJ dysfunction were evaluated. Patients with TMJ dysfunction had symptoms of jaw pain, clicking and popping with opening the mouth, varying degrees of limitation of mouth opening, mild to moderate muscle pain and tenderness, and mild to moderate TMJ arthralgia. The control group demonstrated a high level of thermal symmetry over the TMJ region, consistent with earlier studies.[6] However, the patient group had a low level of thermal symmetry and an increased area ΔT value of 0.4 °C.[2] Using the area ΔT values obtained from the thermograms, the authors were able to select the control group from the patient group with 85% sensitivity (17 of 20), and 92% specificity (22 of 24), and an overall accuracy of 89% (39 of 44), when all 44 subjects were included. The results of this study suggest that thermography can be a useful method of diagnosing mild to moderate TMJ disorders.

The intra-articular temperature of the TMJ in patients with rheumatoid arthritis with unilateral involvement has been found to be reduced on the symptomatic side. This asymmetry has been found to be associated with joint crepitus and structural damage of the joint.[9] Appelgren et al.[1] evaluated the relationship between the intra-articular temperature of the arthritic TMJ and the joint fluid concentration of neuropeptide Y-like immunoreactivity (NPY-LI). The authors found that the intra-articular temperature was inversely correlated with the joint fluid concentration of NPY-LI. They felt that the decrease in intra-articular temperature was an early sign of pathophysiology in the TMJ, caused by the intrasynovial release of NPY-LI, which is a strong vasoconstrictor.

Thermography and Headaches

Thermography has also been used in the evaluation of headaches, including migraines, trigeminal neuralgia, sinusitis, glossopharyngeal neuralgia, temporal arteritis, cluster headaches, and post-herpetic neuralgias.[4,10,15]

A 1984 study by Drummond and Lance[4] used thermography to study the relationship between clinical features of headaches and changes in the extracranial circulation. They assessed extracranial vascular changes thermographically and by the change in headache intensity when pressure was applied over the superficial temporal and common carotid arteries during 209 separate headache attacks. The authors found that in patients with unilateral headaches, an increased heat loss from the affected frontotemporal region was observed most frequently. This heat loss was temporarily relieved by compression of the superficial temporal artery. In patients whose headache symptoms resolved, the thermogram returned to a normal, symmetrical pattern. The authors concluded that while thermography was useful in detecting heat loss, they were unable to clearly identify specific vascular changes that allowed them to differentiate between migraine and tension headaches.

In 1985, Kudrow[10] used thermography to study facial temperature changes in cluster headaches. A large group of cluster headache patients was compared to three non-cluster headache groups: classical migraine, hemiplegic migraine, and common migraine. In the patients with cluster headaches, classical migraines, and hemiplegic migraines, 67% to 75% had an asymmetric, ipsilaterally decreased supraorbital temperature distribution. However, the common migraine patients had a significantly less frequent occurrence of this temperature distribution. In addition, the author identified a distinctive facial temperature pattern felt to be pathognomonic for cluster headaches. This pattern consisted of two warm vertical lines that connected higher up in the forehead and was labeled a "chai" pattern. Doppler studies revealed that this pattern was formed by the supraorbital and superficial temporal arteries and their anastomotic vessels.

Swerdlow and Deiter[15] in 1987 used thermography to evaluate the effect of hyperoxia on vascular headache patients and non-headache individuals. One hundred percent oxygen was given to 30 migraine and mixed headache patients and 20 non-headache controls at 10 liters/minute for 5 minutes. The authors found that while the control subjects had reduced facial temperatures post-inhalation of 100% oxygen, 47% of the headache patients had a 0.5 to 1.0 °C increase in temperature. This statistically significant difference in response (P<.01) continued during the post-inhalation phase. The type of headache did not affect results.

Clinically, the hyperoxia did not improve symptoms in headache patients, nor was the response to hyperoxia different in those patients actively experiencing a headache. The authors felt that this study supports the theory of vascular differences between headache and non-headache individuals.

Other Conditions

Facial Herpes Zoster

Facial Herpes Zoster was evaluated using thermography by Sekiguchi et al. in 1994.[13] They used thermography to obtain objective data on the progression of herpes zoster. The authors found that the acute phase of facial herpes zoster was hyperthermic. In the chronic phase, the images became isothermal.

Prostaglandin E Ointment

Takayama et al.[16] evaluated the efficacy and safety of PGE (Prostaglandin E) ointment in eleven patients with chronic orofacial pain. Thermograms were taken before and after the application of PGE ointment to the painful region. The post-treatment thermographic images revealed an increased temperature at the applied site in 8 of 11 patients. This correlated with improvement in the symptoms. Patients in the study who benefited from the PGE ointment had diagnoses of atypical facial pain, posttraumatic facial pain, trigeminal paresthesias, postherpetic neuralgia, radiation stomatitis, and reflex sympathetic dystrophy. The authors suggest that the vasodilating effect of PGE is related to the pain relief.

Thermography and Facial Paralysis

Zhao et al.[18] in 2005 used thermography to determine the effectiveness of different treatment modalities in treating facial paralysis. Patients were divided into three groups and received either direct acupuncture to the stellate ganglion, a stellate ganglion block, or a combination of the two methods. All groups showed improved facial movement and increased temperature on thermography post treatment, with the combined group having the best results. The combined group had a 99.4% improvement, the stellate ganglion block group a 71% improvement, and the acupuncture group a 78.1% improvement. The authors found that the temperature increase was faster and better maintained in the combined treatment group.

Facial thermography has many applications for diagnostic and treatment purposes as well as for studying physiological function. As a safe, noninvasive modality, it can play an important role in the diagnosis and treatment of orofacial pain and has great potential for further use in Dentistry and Medicine.

References

1. Appelgren A, Appelgren B, Kopp S, Lundeberg T, and Theodorsson E. Relation between the intra-articular temperature of the temporomandibular joint and the presence of neuropeptide Y-like immunoreactivity in the joint fluid. A clinical study. *Acta Odontol Scand,* 1993;51(1):1-8.
2. Canavan D and Gratt BM. Electronic thermography for the assessment of mild and moderate temporomandibular joint dysfunction. *Oral Surg Oral Med Oral Pathol Oral Radiol Endod,* 1995;79(6):778-86.
3. Crandell CE and Hill RP. Thermography in Dentistry: A Pilot Study. *Oral Surg Oral Med Oral Pathol,* 1996;21(3):316-20.
4. Drummond PD and Lance JW. Facial temperature in migraine, tension-vascular and tension headache. *Cephalalgia,* 1984 Sep;4(3):149-58.
5. Gratt BM and Sickles EA. Electronic Facial Thermography: An Analysis of Asymptomatic Adult Subjects. *J Orofac Pain,* 1995;9(3):255-265.
6. Gratt BM and Sickles EA. Thermographic characterization of the asymptomatic TMJ. *J Orofacial Pain,* 1993;7:7-14.
7. Gratt BM, Graff-Radford SB, Shetty V, Solberg WK, and Sickles EA. A six-year clinical assessment of electronic facial thermography. *Dentomaxillofac Rad,* 1996;25:247-255.
8. Gratt BM, Sickles EA, Wexler CE, and Ross JB. Thermographic characterization of internal derangement of the temporomandibular joint. *J Orofac Pain,* 1994;8(2):197-206.
9. Kopp S, Akerman S, and Nilner M. Short-term effects of intra-articular sodium hyaluronate, glucocorticoid, and saline injections on rheumatoid arthritis of the temporomandibular joint. *J Craniomandib Disord Facial Oral Pain,* 1991;5:231-8.
10. Kudrow, LA. Distinctive Facial Thermographic Pattern in Cluster Headache: The "Chai" Sign. *Headache,* 1985;25(1):33-6.

11. Pogrel MA, Erbez G, Taylor R, and Dodson T. Liquid crystal thermography as a diagnostic aid and objective monitor for TMJ dysfunction and myogenic facial pain. *J Craniomandib Disord*, 1989;3(2):65-70.

12. Pogrel MA, McNeill C, and Kim JM. The assessment of trapezius muscle symptoms of patients with temporomandibular disorders by the use of liquid crystal thermography. *Oral Surg Oral Med Oral Pathol Oral Radiol Endod*, 1996;82(2):145-51.

13. Sekiguchi T, Otsuka S, Sato H. Face Scale (Lorish) as a pain measurement in clinical course of patients with herpes zoster. *Jpn Pharmacol Ther*, 1994;22(7):361-71.

14. Soffin CB, Morse D, Seltzer S, Lapayowker M. Thermography and Oral Inflammatory Conditions. *Oral Surg Oral Med Oral Pathol*, 1983;56(3):256-62.

15. Swerdlow B and Dieter JN. The thermographically observed effects of hyperoxia on vascular headache patients and nonheadache individuals. *Headache*, 1987 Nov;27(10):533-9.

16. Takayama H, Seo K, Miura K, Tanaka Y, Kobayashi Y, and Someya G. Treatment of chronic orofacial pain with prostaglandin E ointment. *J Jpn Dent Soc Anesthes*, 1996:24(3);518-522.

17. Weinstein SA. Standards for neuromuscular thermographic examination. *Mod Med: Suppl.* 1986;1:5-7.

18. Zhao Y, He L, and Zhang O-H. Effectiveness of three different treatments for peripheral facial paralysis. *Chin J Clin Rehabil*, 2005:9(29);41-3.

Thermography and Golf: Implications for Sports Medicine Applications

Edwin F. Richter III, MD

Conventional radiologic imaging plays a major role in diagnosing sports injuries, but in most situations the patient must remain stationary for the imaging to be performed. This limitation is typically not a problem when assessing static anatomic changes, but may hinder assessment of dynamic problems that occur primarily when the athlete is in motion. Electrodiagnostic testing may provide valuable physiologic data about muscle or nerve pathology, but it does not assess other types of soft tissue pathology. Moreover, the invasive nature of the procedure may limit patient compliance, and it is not conducive to studies in which athletes perform the motions of their sports. This raises the question of how to better measure physiologic changes that occur during performance of athletic activities.

Benefits of thermography evaluation

Kastberger & Stachl (2003)[8] outlined several advantages of thermography (computerized infrared imaging or CII) in biological applications. These included rapid imaging, which allows effective measurement of moving targets with no need for direct contact with the subject. Since no direct contact is required, there is no interference with the natural motions performed by athletes.

In both clinical and research settings, there is often interest in obtaining additional data after an intervention has been performed. Infrared imaging does not require exposure of subjects to radiation, and there is no need to obtain intravenous access or administer contrast materials. This allows repeated follow-up thermography without risk of adverse effects from the procedure. Since the procedure is not painful or invasive, it does not present the disincentives to participation that are associated with some other diagnostic testing modalities.

Golf is a non-contact sport open to participants across a wide range of ages. Although perceived as a safe activity requiring lesser levels of fitness than some other sports, there is still significant potential for injury. Aside from the obvious risk of being struck by a ball or club, there are risks of injury from repetitive stress.

During athletic activities, forces are exerted on biological structures. Repeated application of loading forces may lead to microtrauma of affected tissues.[1] The practical question for clinicians is how to assess such pathology, as the diagnostic approach to acute major trauma is not particularly applicable in these situations. If the reported complaints indicate gradual progressive onset of symptoms, the clinician may inquire in detail about the activities that exacerbate symptoms.

Studies

The majority of golf musculoskeletal injuries involve overuse syndromes. Gosheger et al. (2003)[5] analyzed injury data from 703 golfers and found that 82.6% of reported injuries involved overuse versus 17.4% from single episodes of trauma. Back, shoulder, elbow and wrist were common sites of injury.

Biomechanical differences have been found between golfers with and without low back pain. Lindsay & Horton (2002)[9] found distinct differences in swing mechanics between these two groups in a study of 6 golfers with low back pain and 6 without low back pain. Vad et al. (2004)[13] studied 42 consecutive male professional golfers, of whom 33% had history of low back pain greater than 2 weeks duration affecting play within the past year. They found a statistically significant correlation between histories of low back pain and decreased lead hip internal rotation on range of motion assessment.

Case reports have highlighted the potential for work on swing mechanics and exercise to help with low back pain[6] and elbow pain.[7] These reports also stressed the challenges involved in treating patients who are eager to continue participation in the activity that has caused their injuries.

Among the methods of teaching golf swings is a system called Natural Golf™, designed to emulate the swing mechanics of the late pro-

The Natural Golf™ (NG) swing method includes a wider stance than that used in conventional golf (CG) swings, and requires less rotation at the hips.

fessional golfer Moe Norman. The Natural Golf™ (NG) swing method includes a wider stance than that used in conventional golf (CG) swings, and requires less rotation at the hips.[10] While the primary purpose of this method is to improve performance, this biomechanical difference represents a potential advantage for golfers who have experienced musculoskeletal pain with their usual golf swing.

Prospective study of golfers trained in different swing techniques would be the most compelling way to evaluate the relative benefits in terms of subsequent rates of musculoskeletal injuries. Before embarking on such an ambitious endeavor, however, it would be useful to know if there is supporting evidence to indicate that an alternative swing method would have a meaningful impact.

Infrared imaging has played a role in clinical evaluation for many years. It has long been established that the normal state of the human body is bilateral thermal symmetry. Feldman & Nickoloff (1984)[3] and Uematsu et al. (1988)[12] demonstrated normal values two decades ago. Thermographic imaging systems, although more familiar to the public in the context of military applications, have been used in clinical studies. Pochaczevsky (1987)[11] and Garagiola and Giani (1990)[4] explored sports medicine applications with earlier versions of the technology used currently.

Measuring the effects of repetitive activities

Prior work in the Kathryn Walter Stein Chronic Pain Laboratory, using thermography technology, has confirmed the baseline heat emission symmetry of normal individuals,[2] and shown that at rest they demonstrate symmetrical cooling.[14] Repetitive activities can be studied to evaluate how they interfere with this process.

Thermography, which is a noninvasive technology, allows an opportunity to study the effects of different golf swings on the heat emission patterns of various areas of the body. Since the apparatus does not touch the subject, it will not interfere with the performance of the golf swing.

The ability of a certified Natural Golf™ instructor,* with a long history as a golf professional prior to exposure to this method, to consistently and repeatedly swing a golf club with either a CG or NG swing, was used to explore the potential application of thermography technology for dynamic physiologic assessment of golfers.

After the upper body was appropriately disrobed and allowed to equilibrate to the ambient temperature of the laboratory, a set of baseline images was obtained. Subsequently a series of 100 CG swings were taken with a standard 9-iron, with repeat imaging done after every 20 swings. Aside from the brief interruptions for imaging, the pace was more consistent with that of a golfer hitting balls at a driving range than that of a golfer playing a game.

The following day the same process was repeated with a series of 100 NG swings with the same club, with baseline imaging and imaging every 20 swings.

---

*   Many thanks to Ken Martin, PGA Professional, Director of Instruction, Natural Golf™, and Executive Director of the Moe Norman Golf Academy, for his assistance with this project.

At the time of assessment, no pain was reported by the subject. A past history of some left elbow pain while golfing (prior to adopting the NG swing) was reported.

The results of the CG swings are seen in Figures 1 and 2, and those of the NG swings are seen in Figures 3 and 4. The overall trend, as expected, was for gradual cooling of exposed areas of the body. Exertion from swinging a golf club at a comfortable pace was generally not sufficient to reverse this cooling trend. With the CG swing, however, the left lateral elbow was noted to actually increase in temperature by 0.12 °C, while

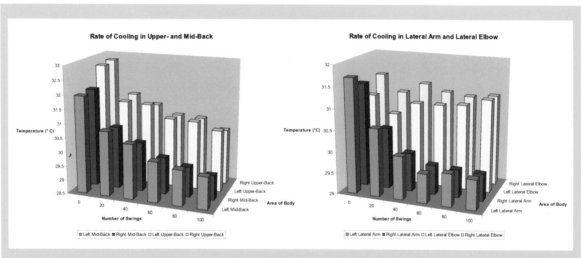

Figure 1. *CG swing in mid and upper back*

Figure 2. *CG swing in lateral arm and lateral elbow.*

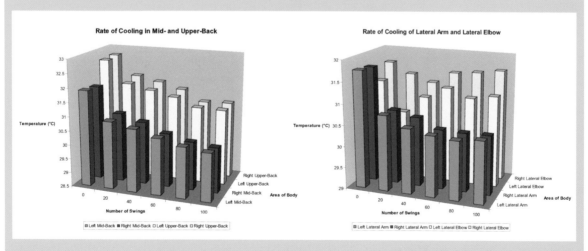

Figure 3. *NG swing in mid and upper back.*

Figure 4. *NG swing in lateral arm and lateral elbow.*

the right side showed cooling of 0.31 ºC. In contrast, with the NG swing the left and right elbows cooled 0.15 ºC and 0.02 ºC respectively.

With the NG swing, the rates of cooling over the back were quite symmetrical. With the CG swing, a trend toward asymmetrical cooling was seen. While the difference of 0.30 ºC degrees in the rate of cooling at 100 swings in the mid-back region would not signal a pathologic abnormality, one must note that 100 swings at a comfortable pace by a trained professional was not an exceptionally demanding workload.

These data have intriguing implications. An individual not currently in pain showed a trend toward asymmetrical heat emission responses with his CG swing and symmetrical responses with his NG swing. This suggests that biomechanical differences in swing technique are leading to physiological differences in relevant body areas. Further exploration with different applications of stresses, such as variations in golf club length and weight, or different numbers of repetitions, might expand upon this area of investigation.

Opportunities for future thermography testing

Imaging actively symptomatic individuals may reveal different types of information. Repeat imaging after intervention in symptomatic individuals might show changes from the initial imaging, perhaps even before symptoms resolve. Such information might be useful to a clinician trying to decide whether to continue with a particular course of treatment, since a short period of rest, short course of therapy, and/or change in swing mechanics might not immediately lead to resolution of symptoms.

A golfer's swing is just one of the many repetitive motions performed in sports. Studying a person in motion without interfering with performance of the motion can be a challenge. Attachment of equipment to an athlete's body to perform measurements during motion may at least subtly interfere with performance. The technology used to obtain measurements should provide the least possible interference with normal performance, but still must obtain data accurately and rapidly.

The testing environment should be sufficiently adaptable to allow the closest possible re-creation of the actual sports activity.

If an athlete uses equipment while performing an activity such as swinging a baseball bat, one would like to have them use the actual piece of equipment during testing to most accurately reproduce the activity being studied. In this situation, swinging a baseball bat, tennis racket, or golf club could produce greater forces on parts of the athlete's body than if the athlete were to try to simulate the motion of a swing while empty-handed. The testing environment should be sufficiently adaptable to allow the closest possible re-creation of the actual sports activity.

Summary

Thermography has many evident advantages as a diagnostic and research modality in sports medicine. As a completely noninvasive modality that does not expose the subject to any radiation from the testing apparatus, this technology allows the athlete to safely perform the activity that exacerbates their symptoms. Asking about exacerbating activities is a routine part of clinical evaluation, but this procedure can now provide physiologic data about the patient before, during, and after performance of such activities. If part of the treatment plan involves retraining, such as alteration of swing or throwing mechanics, re-testing can be done to measure alterations in the physiologic data, without exposing the patient to harm from the testing technology. While one would not expect thermography to replace other testing modalities entirely, it can have widespread useful applications in sports medicine.

References

1. Barry NN and McGuire JL. Overuse syndromes in adult athletes. *Rheum Dis Clin North Am,* 1996;22(3):515-30.
2. Cabrera IN, Wu SSH, Haas F, and Lee MHM. Quantitative Analysis of Cutaneous Temperature Utilizing Computerized Infrared Imaging in Patients with Pain. *AAP Annual Meeting Abstracts, Am J Phys Med Rehabil,* 2001;80(4):310.
3. Feldman F and Nickoloff E. Normal thermographic standards in the cervical and upper extremities. *Skeletal Radiol,* 1984;12:235-49.
4. Garagiola U and Giani E. Use of telethermography in the management of sports injuries. *Sports Med,* 1990;10:267-72.
5. Gosheger G, Liem D, Ludwig K, Greshake O, and Winkelmann W. Injuries and overuse syndromes in golf. *Am J Sports Med,* 2003;31(3):438-43.

6. Grimshaw PN and Burden AM. Reduction of low back pain in a professional golfer. *Med Sci Sports Exerc,* 2000;32(10):1667-73.

7. Grimshaw P, Giles A, Tong R, and Grimmer K. Lower back and elbow injuries in golf. *Sports Med,* 2002;32(10):655-66.

8. Kastberger G and Stachl R. Infrared imaging technology and biological applications. *Behav Res Methods Instrum Comput,* 2003;35(3):429-39.

9. Lindsay D and Horton J. Comparison of spine motion in elite golfers with and without low back pain. *J Sports Sci,* 2002;20(8):599-605.

10. *Natural Golfer Magazine,* Spring 2005 Guide to Better Golf, "The four fundamentals," http://www.naturalgolf.com/ClubHouseMagazine.aspx?vol=007&iss=001&art=story2_4fundamentals.

11. Pochaczevsky R. Thermography in posttraumatic pain. *Am J Sports Med,* 1987;15(3):243-50.

12. Uematsu S, Edwin DH, Jankel WR, et al. Quantification of thermal asymmetry, Part I: Normal values and reproducibility. *J Neurosurg,* 1988;69:552-5.

13. Vad VB, Bhat AL, Basrai D, Gebeh A, Aspergren DD, and Andrews JR. Low back pain in professional golfers: the role of associated hip and low back range-of-motion deficits. *Am J Sports Med,* 2004;32(2):494-7.

14. Wu SSH, Richter EF, Cohen JM, Rosenblum JA, and Lee MHM. Physiologic response to acupuncture as evidenced by computerized infrared imaging [Abstract]. *Arch Phys Med Rehabil,* 1996;77:960.

Use of Acupuncture and Thermography in Modern Medicine

Jay Rosenblum, MD
Sam S.H. Wu, MD, MA, MPH, MBA
Izumi Nomura Cabrera, BM, MA, MD
Mathew H.M. Lee, MD, MPH

Acupuncture has been used to treat pain effectively for several millenniums. The proposed mechanism of action for acupuncture involves the sympathetic nervous system as well as the opioid and non-opioid systems. Trigger point injections and transcutaneous electrical stimulation are the Western medical equivalents of acupuncture.

Pain is often difficult to document because of the dearth of objective tools available to physicians. Many chronic pain patients have sympathetically maintained pain. Thermography is a non-invasive procedure that can objectively record the surface skin temperature changes that are controlled by the sympathetic nervous system. Thermography can be used to objectively document the patients' pain complaints and some of the effects of acupuncture treatments for pain control.

Acupuncture, an ancient healing art, can now be documented with the use of modern technology. Acupuncture has been said to originate in prehistoric ancient times in China as a healing art.[15] It was first used in this country in the early nineteenth century and then temporarily dismissed, until Dr. Osler reintroduced it in his textbook of medicine, *Principles and Practice of Medicine,* in 1892.[21] Acupuncture remained as a subject in all subsequent editions of this book up until 1944, when it was again disregarded.

The use of acupuncture reemerged with the political thaw between the United States and China. In 1971, the use of acupuncture was demonstrated to American physicians during operative procedures.[13]

However, medical skepticism remained regarding the validity of acupuncture use for pain relief. Nevertheless, in the past twenty years there has been a growing acceptance of this mode of analgesia. Acupuncture, electroacupuncture, and acupressure have been known to be very beneficial in a wide variety of pain syndromes.[4,14,20] Still, many Western physicians feel that acupuncture is an empiric treatment with unproven scientific basis.[8,26]

Mechanism of Acupuncture

The mechanism of acupuncture, according to Traditional Chinese Medicine (TCM), is a holistic system of *Yin* and *Yang*. Yin and Yang are said to be two opposing forces in the body, and sickness results from disturbances in the balance between Yin and Yang. This idea is consistent with the modern equivalent of homeostasis and equilibrium in the internal environment.

In TCM, a second mechanism said to influence disease is the state of *Qi* in the body. Stagnation of Qi along the meridians is thought to cause illness. Acupuncture therapy is used to bring Yin and Yang into balance and to release the stagnation of Qi. The acupuncturist thus must be knowledgeable in the meridian points and be aware of the stimulation necessary to achieve this balance.[14]

Stimulation at the appropriate meridian point could result from needling alone, needling with the addition of electricity, or acupressure alone. The modern equivalent of electro-acupuncture is Transcutaneous Electrical Nerve Stimulation (TENS), which first burst on the scene in the late 1970s. It was originally introduced with diagrams to show the appropriate stimulation points drawn along the human figure. These TENS diagrams appeared to be similar to acupuncture meridians.[20]

In an interesting study from the department of medicine at the University of Erlangen-Neurenberg Germany, Drs. Lux et al. performed a randomized study of the effect of acupuncture on eight healthy volunteers. Gastric acid secretion was investigated.[17] Electro-acupuncture at the appropriate acupuncture point was found to reduce gastric acids as compared to a control period without acupuncture. In addition, TENS was also found to reduce gastric acid secretion. Sham acupuncture, classic needle acupuncture, and laser acupuncture had no effect on the gastric acid secretion. These studies, which were performed on healthy

> The mechanism of acupuncture, according to Traditional Chinese Medicine (TCM), is a holistic system of *Yin* and *Yang*.

volunteers, suggest the intimate relationship between transcutaneous nerve stimulation and electro-acupuncture. It also indicates that for electro-acupuncture to be successful, it must be performed at defined points to reduce gastric secretion.

Acupressure has also been found to be similar to trigger point areas. Melzack et al. found over 70% correspondence between trigger points and acupuncture points. They concluded that this "close correlation suggests that trigger points and acupuncture points for pain, though discovered independently and labeled differently, represent the same phenomenon and can be explained in terms of the underlying neural mechanisms."[18,19]

In the late 1940s, trigger point injections became popular in the United States. Travell et al. demonstrated that injections with Procaine, plain physiological saline, or "dry needling" were all effective for the alleviation of pain in the trigger point areas. Travell and Simons reported[34] that "precise dry needling of trigger points without injecting any solution approaches, but does not quite equal, the therapeutic effectiveness of injecting Procaine into the trigger point."

In the 1980s, Frost et al. in Denmark published a study comparing mepivacaine injections to saline injections for myofascial pain. They demonstrated that physiological saline injections were statistically more effective than mepivacaine injections and attributed the effectiveness to an irritation from the needling. They concluded "there is much to suggest that trigger point injection therapy for myofascial pain is one form of acupuncture."[6]

The currently accepted scientific theory of the mechanism of acupuncture dates back to the early 1970s.[14] Acupuncture analgesia was a result of either a neuropharmacological or humoral mechanism. Chinese researchers demonstrated that the analgesic effect of acupuncture can be transmitted between animals, either through cross-circulation techniques or through infusion of cerebral ventricular fluid transmission.[27]

Pomeranz et al.,[23,24,25] reported three important observations from animal experiments: 1) an analgesic effect could not be produced from sham acupuncture, 2) the analgesic effect could be eliminated by the ablation of the pituitary gland, and 3) acupuncture analgesia could be eliminated by naloxone. They also pointed out that the adrenal and pituitary glands had mutually reversing influences on acupuncture analgesia. They[2] postulated that the body produces an endogenous substance called endorphin. Endorphin is a group of naturally occurring chemicals involved in analgesia.

The endogenous opioid system is activated by high intensity, low frequency (2–4 Hz) electro-acupuncture. This type of electro-acupuncture tends to result in a long lasting cumulative analgesia that is slow in onset and can be abolished by naloxone.

The non-opioid system plays an equally important role in acupuncture analgesia. Serotonin, atropine and L-tryptophan have been implicated in acupuncture analgesia. The non-opioid system is generally activated by low intensity, high frequency electro-acupuncture. This type of electro-acupuncture tends to result in a short-lasting, noncumulative analgesia that is rapid in onset.

Studies performed at the University of Western Ontario investigated this phenomenon.[32] Scudds et al. examined the effects of two modes of TENS, compared with a control condition, on the skin temperature of the hand and finger of 24 asymptomatic subjects. All subjects participated in a 4-Hz TENS session, a 100-Hz TENS session, and a control (no TENS) session. Hand temperature was measured with the use of infrared thermography. The results revealed that the mean hand temperature after low-frequency TENS was 1.69 ºC warmer than the mean hand temperature following high-frequency TENS and 1.60 ºC warmer than after the control condition. The authors concluded that high-intensity, low-frequency TENS prevents cooling of the hand. This study suggests that the pain specialist and the acupuncturist must be cognizant of the differences in the therapeutic effectiveness of the varying electrical intensity during electro-acupuncture and the TENS units upon the pain processes.

Acupuncture has an effect on the autonomic nervous system as well.[13] This has been documented by changes in skin temperature and cardiovascular systems, along with pain relief in patients undergoing acupuncture therapy.

Acupuncture, electro-acupuncture, or acupressure all appear to have a beneficial impact on patients with pain syndromes. Most physicians are more comfortable in naming these procedures as a TENS unit, trigger points injection, etc., rather than an "acupuncture-like" procedure. The salient feature to the patient, however, is that there is a relief of pain from these procedures.

The problem that faces the acupuncturist is that on certain individuals, acupuncture may not afford any relief. It is estimated that only 5% of patients in an initial session will experience permanent pain relief. It is known that the clinical effects of acupuncture are cumulative and that the pain is only gradually reduced during the course of treatment. The acupuncturist is often faced with the question of whether a patient will respond to acupuncture and, if so, how many treatments are necessary before a therapeutic plateau is reached. Indeed, The New York Society of Acupuncture for Physicians and Dentists advocates a course of at least six treatments

Infrared imaging is not a picture of pain, but rather a picture of the vascular system manifestations in the body in response to pain.

in order to determine its effectiveness. The Society also indicates that periodic boosters may be necessary to sustain the continued response.[16]

It is estimated that 60–75% of chronic pain patients will benefit from acupuncture. The question facing the acupuncturist is who represents the other 25–40%.[20] Infrared imaging thermography is considered an ideal diagnostic instrument to answer this question.

Measuring the Effects of Acupuncture

Acupuncture in conjunction with thermography would be most cost-effective. The acupuncturist would not have to rely on empirical methods to determine the number of necessary acupuncture treatments. Up until the present time, a positive response to acupuncture treatment has been subjective. The "subjective" conclusions used a yardstick based on two observations: 1) the need for decreased medication and 2) the ability to increase activities of daily living. This yardstick was recorded by both the patient and physician. Infrared imaging could now offer an objective measurement and eliminate the population of non-acupuncture respondents. Unfortunately, what appears to be the ideal marriage for diagnosis and therapy is not universally acknowledged. This is in spite of the multiple medical articles that have appeared in the literature documenting the use of thermography[28,29,30.31,33] in a variety of disease disorders.

Infrared imaging is a diagnostic procedure that measures infrared energy emitted by the skin. Skin temperature, in turn, is a reflection of the cutaneous blood flow under the control of the autonomic nervous system. These measurements are then expressed in the form of thermal images. Pain fibers also have a relationship with the autonomic nervous system. It is this interrelationship that allows thermography to be of diagnostic value. Infrared imaging is not a picture of pain, but rather a picture of the vascular system manifestations in the body in response to pain.

One should be aware that the cutaneous thermal distribution on thermography is not identical to the spinal dermatomal charts with which physicians are familiar; instead, it is more in keeping with the distribution of the cutaneous autonomic blood vessels. Infrared imagers refer to this as a thermatome, which is anatomically different from the dermatome.

With a well grounded knowledge of thermography, infrared imagers in most cases can now objectively map out the thermotone area. It has been found that areas of heat asymmetry in corresponding thermatomes will often be the area of a patient's pain symptoms. This objective documentation assists the physician's search for the cause based on history, physical examination, and other diagnostic studies.

As far back as 1963 in The Journal of the American Medical Association, Barnes[1] concluded that clinical thermography is a good index for sympathetic activity as evidenced by qualitative and quantitative changes of the cutaneous temperature. Since that time, objectivity of thermography has been well documented in multiple national and international articles.[1,3,5,9,10,30,35–37] On all accounts, there seems to be an agreement that thermography may provide information about the cutaneous temperatures, which in turn can be useful in characterizing sympathetically maintained diseases as well as other autonomic neuropathies.

The problem that the infrared imager faces is that thermography has been controversial even among infrared imagers themselves. Dr. Clewell[3] surveyed 257 experienced infrared imagers as to the value of thermography. The physicians felt that thermography was of great diagnostic value for RSD and sympathetically maintained pain, but were less than enthusiastic for its use in entrapment neuropathies, fibromyalgia, and hysteria.

It has been well accepted that the mechanism of reflex sympathetic dystrophy, a neuropathic pain syndrome, has an association with sympathetic vasoconstriction and vasodilation mechanisms manifested by skin flow abnormalities.[1] Infrared imaging, which documents areas of heat asymmetry corresponding to areas of pain, can be used to monitor the clinical benefit of acupuncture.[13]

Acupuncture, a method of pain management, in conjunction with thermography, a non-invasive diagnostic tool, may prove to be a cost-effective outpatient procedure in the physicians' armament against pain.

> On all accounts, there seems to be an agreement that thermography may provide information about the cutaneous temperatures, which in turn can be useful in characterizing sympathetically maintained diseases as well as other autonomic neuropathies.

References

1. Barnes RB, Gershon-Cohen J. Clinical Thermography. *JAMA*, 1963;185:949-52.

2. Cheng RS and Pomeranz B. Electroacupuncture analgesia could be mediated by at least two pain-relieving mechanisms: endorphin and non-endorphin systems. *Life Sci*, 1979;25:1957-62.

3. Clewell WD. Thermographers Assess Thermography. *Am J Pain Manag*, 1995;5(4):133-135.

4. Edwards BE and Hobbins W. "Pain Management and Thermography," in *Practical Management of Pain*, P. Prithvi Raj, Ed. Mosby-Year Book Inc., St. Louis, MO, 1992.

5. Friedman MS. The Use of Thermography in Sympathetically Maintained Pain. *Iowa Orthop J*, 1994;14:141-7.

6. Frost FA, Jessen B and Siggaard-Anderson J. A control, double-blind comparison of mepivacaine injection versus saline injection for myofascial pain. *Lancet*, 1980;1:499-501.

7. Fung YL. *A Short History of Chinese Philosophy*, The Free Press, New York, 1948.

8. Grabow L. Controlled study of the analgesic effectivity of acupuncture. *Arzneimittelforschung*, 1994;44(4):554-8.

9. Jiang LJ, Ng EY, Yeo AC, Wu S, Pan F, Yau WY, Chen JH, and Yang Y. A perspective on medical infrared imaging. *J Med Eng Technol*, 2005;29(6):256-67.

10. Jones BF. A Reappraisal of the Use of Infrared Thermal Image Analysis in Medicine. *IEEE Transactions on Medical Imaging*, 1998;17:1019-27.

11. Kurvers HA, Jacobs MJ, Beuk RJ, van den Wildenberg FA, Kitslaar PJ, Slaaf DW and Reneman RS. The spinal component to skin blood flow abnormalities in reflex sympathetic dystrophy. *Arch Neurol*, 1996;53(1).

12. Lee MHM and Ernst M. Sympathetic vasomotor changes induced by manual and electrical acupuncture of the Hoku point visualized by thermography. *Pain*, 1985;21:25-33.

13. Lee MHM and Ernst M. The sympatholytic effect of acupuncture as evidenced by thermography: a preliminary report. *Orthop Rev*, 1983;12:67-72.

14. Lee MHM and Liao SJ. "Acupuncture for Pain Management," in *Physiatric Procedures in Clinical Practice*. Hanley & Belfus, Philadelpha, 1995;49.

15. Lee MHM and Liao SJ. "Acupuncture in Physiatry," in Krusen's *Handbook of Physical Medicine and Rehabilitation,* Fourth Edition, F.J. Kottke and F. Lehmann, Eds. W.B. Saunders, Philadelphia, 1990.

16. Lee MHM. "Acupuncture for pain control," in *Pain Control: Practical Aspects of Patient Care,* LC Mark, Ed. Masson Publishers, New York, 1981.

17. Lux G, Hagel J, Backer P, Backer G, Vogl R, Ruppin H, Domschke S, and Domschke W. Acupuncture inhibits vagal gastric acid secretion stimulated by sham feeding in healthy subjects. Department of Medicine A., University of Erlangen-Nuremberg Germany, Gut 1994 Aug; 35 (8): 1026-9, Article Number: UI95011726.

18. Melzack R, Stillwell DM and Fox EJ. Trigger points and acupuncture points for pain: Correlations and implications. *Pain,* 1977;3:3-23.

19. Melzack R. Prolonged relief of pain by brief, intense transcutaneous somatic stimulation. *Pain,* 1975;1:373-5.

20. Ng LKY, Katims JJ, and Lee MHM. "Acupuncture: A Neuromodulation Technique for Pain Control," in *Evaluation and Treatment of Chronic Pain,* Second Edition. Williams & Wilkins, Baltimore, 1992;291.

21. Osler W. *Principles and Practice of Medicine,* Appleton, New York, 1892.

22. People's Republic of China: Acupuncture Anesthesia (movies), 1973.

23. Pomeranz B and Chui D. Naloxone blockade of acupuncture analgesia: endorphin implicated. *Life Sci,* 1976;79:1757-62.

24. Pomeranz B. "Scientific basis of acupuncture," in *Acupuncture: Textbook and Atlas,* G. Stuz and G. Pomeranz, Eds. Springer-Verlag, Berlin, 1987:1-34.

25. Pomeranz BM, Cheng RM, and Law P. Acupuncture reduces electrophysiological and behavioral responses to noxious stimuli: Pituitary is implicated. *Exp Neurol,* 1977;54:172-8.

26. Resch KL and Ernst E. Proving the effectiveness of complementary therapy, Analysis of the literature exemplified by acupuncture. Post-graduate Medical School, University of Exeter/UK., Fortschr Med 1995 Feb 20;113 (5):49-53, Article Number UI95237741.

27. Research Group of Acupuncture Anesthesia, Peking Medical College, Peking. The role of some neurotransmitters of the brain in finger-acupuncture analgesia. *Sci China,* 1974;(B)17:112-3.

28. Rosenblum J, Spielholz N, Lee MHM, and Ma D. Reflex Sympathetic Dystrophy of the Upper Extremity Following Infection and Removal of a Silicone Breast Implant. *J Neurol Orthop Med Surg*, 1992;13(2):131-5.

29. Rosenblum JA, Cohen JM, and Lee MHM. Hyperhidrosis – a case history. *Angiology*, 1993;45(1):61-4.

30. Rosenblum JA. Documentation of Thermographic Objectivity in Pain Syndromes. Academy of Neuromuscular Thermography: *Clin Proc Postgrad Med*, 1986(March);59-61.

31. Rosenblum JA. Femoral & Lateral Femoral Cutaneous Neuropathy: Neurological Complications of Hysterectomy. *J Neurol Orthop Med Surg*, 1988;9(2):143-8.

32. Scudds RJ, Helewa A, and Scudds RA. The effects of transcutaneous electrical nerve stimulation on skin temperature in asymptomatic subjects. Department of Physical Therapy, University of Western Ontario. *Phys Ther*, 1995;75(7):621-8A, Article Number UI95327735.

33. Spielholz NI, Rosenblum JA, Lee MHM and Geisel LC. Case Report – Unilateral Leg Pain in a Drug-Abuser Following Ipsilateral Rhabdomyolyosis and Peripheral Nerve Injury. *Am J Pain Manag*, 1993;3(2):57-9.

34. Travell JG and Simons DG. *Myofascial Pain and Dysfunction: The Trigger Point Manual*. Williams & Wilkins, Baltimore, 1983, pp. 20, 76.

35. Uematsu S, Edwin DH, Jankel WR, et al. Quantification of thermal asymmetry Part I: Normal values and reproducibility. *J Neurosurg*, 1988;69:552-5.

36. Uematsu S, Edwin DH, Jankel WR, et al. Quantification of thermal asymmetry Part II: Application in low-back pain and sciatica. *J Neurosurg*, 1988;69:556-61.

37. Uematsu S. Thermographic imaging of cutaneous sensory segments in patients with peripheral nerve injury. *J Neurosurg*, 1985;62:716-20.

Acupuncture and Thermography — Clinical Studies

Jeffrey M. Cohen, MD
Mathew H.M. Lee, MD, MPH
Laura Downing, BS

Early studies in the 1970s on the role of the sympathetic nervous system in acupuncture analgesia provided conflicting results. Some authors found an increase in sympathetic activity following acupuncture in dogs,[7] while others reported a decrease in sympathetic tone, resulting in peripheral vasodilatation.[8,9] Omura in 1976 reported that acupuncture resulted in a transient increase in sympathetic activity followed by a long-lasting decrease.[14] The discrepancies in these findings arise not only from the difficulty of monitoring sympathetic activity, but also from the different types of acupuncture used by the researchers. With the advent of thermography in the late 1950s, however, clinical researchers had a simple, reliable, and sensitive tool for assessing peripheral sympathetic activity through measurements of the surface skin temperature.

Skin temperature is a function of blood perfusion, which is controlled by sympathetic nervous system activity. By assessing changes in skin temperature, the researcher can indirectly assess changes in sympathetic activity. An increase in sympathetic vasomotor tone induces vasoconstriction, thus decreasing skin temperature. Conversely, a decrease in sympathetic vasomotor tone induces vasodilatation and an increase in skin temperature. Thermography permits the researcher to record the body's superficial temperatures, providing an indirect assessment of the underlying sympathetic nervous system activity.

Dr. Mathew Lee pioneered the use of thermography to study the effect of acupuncture on the sympathetic nervous system. In 1983, Lee and Ernst published a preliminary report on the sympatholytic effect of acupuncture, as evidenced by thermography.[10] This pilot study found a skin temperature rise in both hands of three normal subjects after L14 Hoku manual acupuncture, suggesting a sympatholytic effect.

Drs. Ernst and Lee followed up this initial research with two studies designed to measure skin temperature changes in the face, hands, and feet after manual and electrical acupuncture of the 1) L14 Hoku hand point[6] and 2) St.36 Tsusanli knee point.[5] To correct for the discrepancies in acupuncture techniques used in the earlier studies, the authors used both acupuncture modalities (electrical and manual) in these two experiments. There is evidence of a specific analgesic response pattern according to the type of acupuncture used. Generalized analgesia is associated with manual acupuncture,[2,4,13] while segmental analgesia is described with electrical stimulation.[12,15]

Hoku hand point study

The Hoku point study[6] evaluated the effects of acupuncture of the left hand Hoku point on the superficial skin temperature (Tsk) of the face, hands, and feet in 19 normal subjects. Each subject participated in three randomly distributed sessions consisting of a control session, a Hoku manual acupuncture (MA) session, and a Hoku electrical acupuncture (EA) session. The sessions were performed at one week intervals at the same time of day in a temperature-controlled, draft-free room. Tsk of the hands, feet, and face were recorded using infrared thermography. In the treatment sessions, a baseline set of thermographic images was taken, followed by acupuncture to the left Hoku point for 15 minutes. Infrared images of the face, hands, and feet continued to be taken while the acupuncture was being administered. At the end of the acupuncture treatment, thermography images were taken for another 15 minutes, for a total of 30 minutes of imaging. In the control session, subjects sat quietly while thermography images were taken for the full 30 minutes. Mean Tsk of each body segment was calculated by averaging the Tsk of smaller areas: forehead, nose, chin, and cheeks for the face, phalangeal, metacarpal, and carpal areas for the hands, and phalangeal, metatarsal, and tarsal areas for the feet.

The results of this study showed a progressive and significant Tsk decrease during the control condition, reaching -2 °C for the hands and feet and -0.7 °C for the face. Compared to the control condition, MA induced an immediate, generalized warming effect

The results of the Hoku hand point study are significant in that they are in agreement with the normal physiological spatial distribution and temporal fluctuations of Tsk.

for the face, and a delayed, generalized warming effect for the hands and feet. This increase was greatest at the end of the 30 minute session, and was most significant for the face ($p<0.001$), less significant for the hands ($p<0.01$), and not significant for the feet. The Tsk increase represented a sympatholytic effect with a cranio-caudal distribution. EA similarly induced an immediate, warming effect in the face, which became generalized by the end of the session. However, in contrast to the MA findings, EA produced a significant, transient, early Tsk decrease in both the hands ($p<0.05$) and the feet ($p<0.05$). In comparing MA to EA, the only areas of significant change in mean Tsk were in the hands and feet. Both acupuncture modalities induced a mean Tsk increase in the face, but MA yielded a stronger effect. In the hands, MA and EA produced early opposite Tsk changes, with MA increasing hand Tsk while EA decreased hand Tsk.

The results of the Hoku hand point study are significant in that they are in agreement with the normal physiological spatial distribution and temporal fluctuations of Tsk. Tsk is a direct function of the cutaneous blood perfusion, and is not homogenously distributed within the body. Thermal maps from whole body thermograms on resting nude subjects show that the trunk and head present at higher Tsk than the limbs, while the limbs present a positive Tsk gradient from distal to proximal parts.[16] This was reflected in the control condition where the face was 1.5 °C warmer than the hands, and the hands about 1 °C warmer than the feet, following 20 minutes of rest. The Tsk increase following MA also showed this cranio-caudal distribution.

The thermograms of the hands showed no local reaction around the acupuncture needle, and the Tsk changes were symmetrical in their distribution at all times of observation and in all experimental conditions. This suggests the acupuncture produced a central spinal or supra-spinal mediated sympathetic effect, rather than a peripheral effect. Two effects have been proposed to account for the Tsk changes associated with Hoku acupuncture: an initial cooling effect (sympathomimetic) and a generalized long-lasting warming effect (sympatholytic) distributed according to a cranio-caudal gradient. The initial short-lasting cooling sympathetic activation effect appeared in the feet for both MA and EA, and only in the hands in EA. The generalized, long-lasting warming sympathetic inhibitory effect follows this cooling effect, and was stronger with MA than EA. The warming effect predominated in the face area and followed a cranio-caudal distribution. These results are consistent with clinical observations reporting a sensation of well-being and warmth associated with acupuncture treatment.[1] The generalized distribution of the warming effect reached at the end of the session suggests the activation of a central inhibitory system.

The authors concluded that acupuncture of the Hoku point produces two opposite sympathetic effects:

1. An inhibition of sympathetic tone clearly of central origin, although a segmental participation cannot be ruled out.
2. An activation of the sympathetic tone which may reflect either an emotional reaction to acupuncture initiation, or an additional segmental effect specifically related to EA of the Hoku point.

While the Hoku point study confirmed the feasibility of the thermographic technique, it did not permit the researchers to delineate between spinal and central effects of acupuncture, nor did it provide any information as to the specific action of the acupuncture point. Drs. Ernst and Lee decided to continue the experiment by studying the effects of acupuncture on the Tsusanli knee point, which is nonsegmentally related to the Hoku point.

Tsusanli knee point study

The Tsusanli knee point study[5] followed the same experimental procedure as the Hoku point study, except for the area of acupuncture administration. In the treatment sessions, MA or EA was performed for 15 minutes to the Tsusanli knee point.

Drs. Ernst and Lee found the same generalized, long-lasting warming (sympathetic inhibition) effect after MA or EA of the Tsusanli knee point as they did with the Hoku hand point. The warming effect was again greater with MA than EA, and for both types of acupuncture followed the same cranio-caudal distribution, with a maximum effect in the face. The only statistical difference between MA of both points was in the magnitude of the Tsk increase in the face; the Hoku hand point had the strongest effect. As with EA of the Hoku point, they also observed a segmental, short-lasting cooling (sympathetic activation) effect following EA of the Tsusanli point that decreased during the session. As with the Hoku point, the cooling effect was segmentally related to the acupuncture site. The Tsk decrease was observed in the feet following Tsusanli point EA and in the hands following Hoku point EA.

The finding of a central warming sympathetic inhibition in the Tsusanli study confirmed the results of the Hoku study. The similar temporal course and spatial distribution of Tsk changes following stimulation of either the knee or hand point is consistent with the hypothesis of central sympathetic inhibitory system involvement. The somatotopicity of this system is unrelated to the peripheral stimulation site. The Tsusanli knee point study also clarified the mechanism of initial EA sympathetic activation observed in each study. Both Tsk decreases were segmentally related to the acupuncture site, indicating a segmentally related response, not an effect mediated by a generalized emotional arousal. The activation of segmental vasomotor reflexes is most likely responsible for this sympathomimetic effect.[3]

Conclusion

In conclusion, the sympathetic effects of acupuncture observed in these studies are temporally and spatially similar to two separate acupuncture analgesic mechanisms:

1. The long-lasting, generalized sympathetic inhibitory effect observed following MA is correlated with the generalized endogenous opioid analgesia found after MA,[2,4,11] and
2. The short-term, segmentally related sympathetic excitatory effect found with EA correlates to the segmental spinal analgesia produced by transcutaneous electrical stimulation (TENS).[12,15]

The work of Drs. Ernst and Lee shows that changes in skin surface temperature and patterns following acupuncture therapy can be reliably visualized and measured using thermography.

References

1. Academy of Traditional Chinese Medicine. *An Outline of Chinese Acupuncture,* Foreign Language Press, Peking, 1975.
2. Akil H, Mayer DJ, and Liebeskind JC. Antagonism of stimulation produced analgesia by naloxone, a narcotic antagonist. *Science,* 1976;191:961-2.
3. Beachman WS and Perl ER. Characteristics of a spinal sympathetic reflex. *J Physiol* (London), 1964;173:431-48.
4. Cheng RS and Pomeranz B. Electroacupuncture analgesia could be mediated by at least two pain relieving mechanisms: endorphin and non-endorphin systems. *Life Sci,* 1979;25:1957-62.
5. Ernst M and Lee MHM. Sympathetic effects of manual and electrical acupuncture of the Tsusanli knee point: comparison with the Hoku hand point sympathetic effect. *Exp Neurol,* 1986;94:1-10.

6.   Ernst M and Lee MHM. Sympathetic vasomotor changes induced by manual and electrical acupuncture of the Hoku point visualized by thermography. *Pain,* 1985;21:25-33.

7.   Lee DC, Lee MO, and Clifford DH. Cardiovascular effects of acupuncture in anesthetized dogs. *Am J Chin Med,* 1974;2:271-282.

8.   Lee GT. A study of electrical stimulation of acupuncture locus Tsusanli (St36) on mesenteric microcirculation. *Am J Chin Med,* 1974;2:53-56.

9.   Lee MH, Sadove MS, and Kim SI. Liquid crystal thermography in acupuncture therapy. *J Acupunct,* 1976;4:145-148.

10. Lee MHM and Ernst M. The sympatholytic effect of acupuncture as evidenced by thermography: a preliminary report. *Orthop Rev,* 1983;12:62-72.

11. Mayer DJ and Price DD. Central nervous system mechanisms of analgesia. *Pain,* 1976;2:379-404.

12. Melzack R and Wall PD. Pain mechanisms: a new theory. *Science,* 1965;150:971-9.

13. Oliveras JL, Sierralta F, Fardin V, and Besson JM. Implication des systems serotoninergiques dans l'analgesie induite par stimulation électrique de certaines structures du tronc cérébral. *J Physiol* (Paris), 1981;77:473-82.

14. Omura Y. Pathophysiology of acupuncture treatment: effects of acupuncture on cardiovascular and nervous systems. *Acupunct Electrother Res,* 1976;1:51-141.

15. Wall PD and Sweet WH. Temporary abolition of pain in man. *Science,* 1967;155:108-109.

16. Winsor T and Winsor D. "Thermography in cardiovascular disease," in *Medical Thermography, Theory and Clinical Applications,* S. Uematsu, Ed., Brentwood Publishing, Los Angeles, CA, 1976;121-42.

Use of Thermography in Veterinary Medicine

RAM C. PUROHIT, DVM, PhD, DACT [*]

A history and research review of thermography in veterinary medicine has been recently published by Purohit.[21] In the mid 1960s and early 1970s, several studies investigated the value of infrared (IR) thermography in veterinary medicine.[3,5,24] Stromberg[26–28] and Stromberg and Norberg[25] used thermography to diagnose inflammatory changes of the superficial digital flexor tendons in race horses. With thermography, they detected and documented early inflammation of the tendon one to two weeks prior to the detection of lameness using clinical examination. In 1975, Nelson and Osheim[11] documented that soring caused by chemical or mechanical means on the horse's digit could be diagnosed as having a definite abnormal, characteristic IR emission pattern in the affected areas of the limb. Even though thermography at that time became the technique of choice for the detection of soring, normal thermography patterns in horses were not known. Purohit et al.[14] established a protocol for obtaining normal thermographic patterns of the horses' limbs and other parts of the body. This protocol was also used for early detection of acute and chronic inflammatory conditions in horses and other animal species.

In a subsequent study, Purohit and McCoy[13] established normal thermal patterns (temperature and gradients) of the horse, with special attention directed towards thoracic and pelvic limbs. At the same time, Turner et al.[29] investigated the influence of the hair coat and hair clip-

* The author wishes to acknowledge the help of the following faculty for making thermography studies possible: Drs. Bread DeFranco, Dwight Wolfe, Marry A. Williams, Mike D. McCoy, Tracy Turner, Read Hanson, John Schumacher, Robert Carson, L.S. Pablo, Karl Bowman, A.M. Heath, David Pugh, Jay Humburg, S.D. Beckett, Robert Hudson, D.F. Walker, Richard T. Herrick, Jim Brandendmeuhl, G.M. Riddell, J.T. Vaughan, and others who are named as co-authors on several papers referenced in this chapter.

ping. This study demonstrated that the clipped leg was always warmer. After exercise, both clipped and unclipped legs had similar increases in temperature. The thermal patterns and gradients were not altered by clipping and/or exercise.[13,29] This indicated that clipping hair in horses with even hair coats was not necessary for thermographic evaluation. However, in some areas where the hair is long and not uniform, clipping may be required. Recently, concerns related to hair coat, thermographic imaging, and temperature regulation were investigated in llamas exposed to the hot, humid conditions of the southeast United States.[7] These initial studies showed variations in core temperature and differences in the thermoregulatory mechanism among different animal species, demonstrating the need to establish normative standards for each species. A further challenge in veterinary medicine occurs when animal patient care requires outdoor imaging.

Thermography Standards in Veterinary Medicine

Thermography provides an accurate, quantifiable, non-contact, non-invasive measure and map of skin surface temperatures. Skin surface temperatures are variable and change according to blood flow regulation to the skin surface. Thus, the practitioner must be aware of the internal and external influences that alter this dynamic process of skin blood flow and temperature regulation.

The thermographer needs to understand the limitations of their IR system in order to make appropriate interpretations of their data. In some cases, a simple cause-effect relationship was assumed to demonstrate the diagnosis of a disease or syndrome based on thermal responses as captured by thermographic images. However, internal and external factors have a significant effect on the skin surface temperature. Therefore, the use of thermography to evaluate skin surface thermal patterns and gradients requires an understanding of the dynamic changes that occur in blood flow at systemic, peripheral, regional, and local levels.[13,14] To enhance the diagnostic value of thermography, we recommend the following standards for veterinary medical imaging.

1. Environmental factors that interfere with the quality of thermography should be minimized. Room temperature should be maintained between 21 and 26 °C. Slight variations in some cases may be acceptable, but room temperature should always be cooler than the animal's body temperature and free from air drafts.

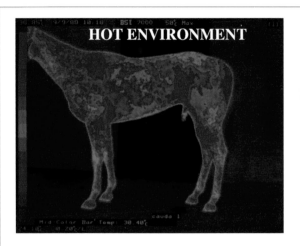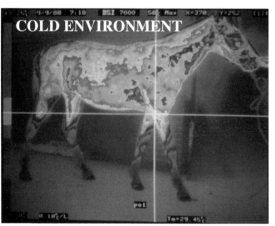

Figure 1: *Environmental factors, such as hot or cold, can interfere with the quality of thermographic images. Thermograms obtained under conditions of extreme variations in temperature may be meaningless as a diagnostic tool.*

2. Thermograms obtained outdoors under conditions of direct air drafts, sunlight, and extreme variations in temperature may provide unreliable thermograms in which thermal patterns are altered. Such observations are meaningless as a diagnostic tool. (Figure 1)

3. When an animal is brought into a temperature-controlled room, it should be equilibrated at least 20 minutes or more, depending on the external temperature from which the animal was transported. Animals transported from extreme hot or cold environments may require up to 60 minutes of equilibration time. Equilibration time is adequate when the thermal temperatures and patterns are consistently maintained over several minutes.

4. Other factors affecting the quality of thermograms are exercise, sweating, body position and angle, body covering, systemic and topical medications, regional and local blocks, sedatives, tranquilizers, anesthetics, vasoactive drugs, skin lesions such as scars, surgically altered areas, etc. As previously stated, the hair coat may be an issue with uneven hair length of a thick coat.

5. It is recommended that thermography (infrared imaging) be performed using an electronic, non-contact, cooled system. The use of long wave detectors is preferable.

It is important to have well-documented, normal thermal patterns and gradients in all species under controlled environments prior to making any claims of detecting pathological conditions.

Dermatome patterns in various animal species

Painful conditions associated with peripheral neurovascular and neuromuscular injuries are easy to confuse with spinal injuries associated with cervical, thoracic, and lumbar-sacral areas.[12,20] Similarly, inflammatory conditions such as osteoarthritis, tendonitis, and other associated conditions may be confused with other neurovascular conditions. Thus, infrared thermography studies were conducted to map the sensory-sympathetic dermatome patterns of cervical, thoracic, and lumbosacral regions in horses.[12,20] The dorsal or ventral spinal nerve(s) were blocked with 0.5% mepivacaine as a local anesthetic. The sensory sympathetic spinal nerve block produced two effects. First, blocking the sympathetic portion of the spinal nerve caused increased thermal patterns and produced sweating of the affected areas. Second, the areas of insensitivity produced by the sensory portion of the block were mapped and compared with the thermal patterns. The areas of insensitivity were found to correlate with the sympathetic innervations.

Clinical cases of cervical area nerve compression provided cooler thermal patterns, away from the site of injuries. In cases of acute injuries, thermal patterns were warmer than normal cases at the site of the injury. Elucidation of dermatomal (thermatome) patterns provided spinal injury locations for the diagnosis of back injuries in horses. Similarly, in a case of a dog with a neck injury (subluxation of atlanto-axis), the diagnosis was determined by abnormal thermal patterns and gradients.

Figures 2–6 illustrate the use of thermography to map out dermatome patterns.

Figure 2: *A horse with C-5, C-6 and C-7 cervical problem, as can be seen in the abnormal sweat pattern. This is the same horse as in Figure 3. Purohit RC, Williams Amy. Department of Large Animal Surgery and Medicine, College of Veterinary Medicine, Auburn University. Unpublished case.*

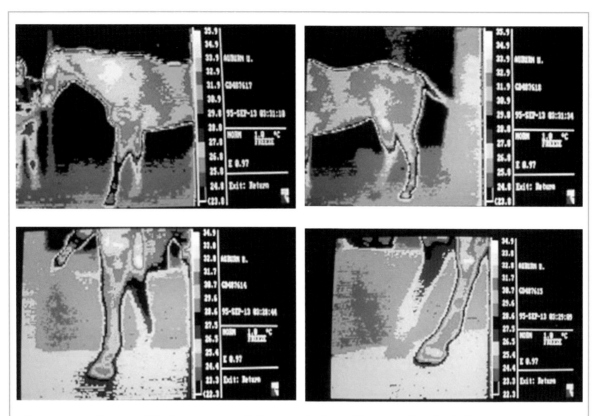

Figure 3: *A horse with C-5, C-6, and C-7 cervical problem causing both front and hind leg lameness on the right. Purohit RC, Williams Amy. Auburn University. Unpublished case.*

Figure 4: *Cervical 6 dermatome in a horse. Purohit RC, DeFranco B. Infrared thermography for the determination of cervical dermatome patterns in the horse. Biomed. Thermology, 1995;188-190.*

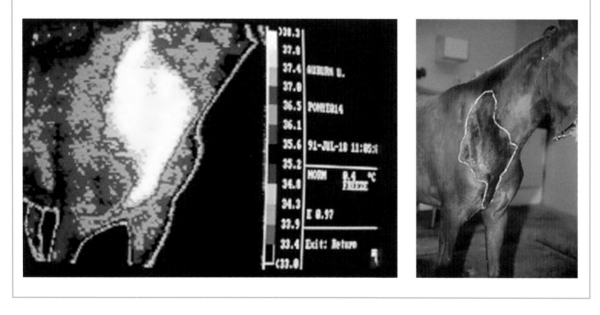

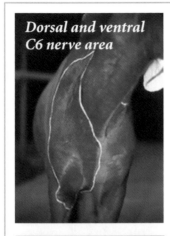

Dorsal and ventral C6 nerve area

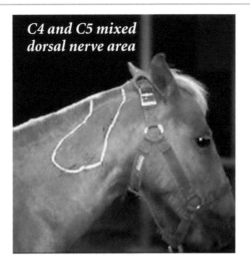

C4 and C5 mixed dorsal nerve area

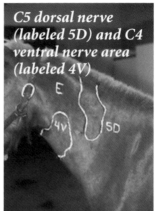

C5 dorsal nerve (labeled 5D) and C4 ventral nerve area (labeled 4V)

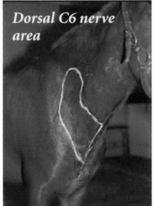

Dorsal C6 nerve area

Figure 5: *Cervical dermatomes in a horse. Purohit RC, DeFranco B. Previously unpublished photos of dermatome studies in horses.*

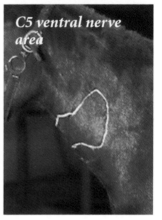

C5 ventral nerve area

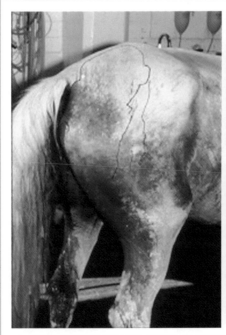

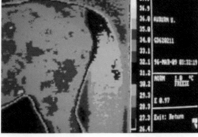

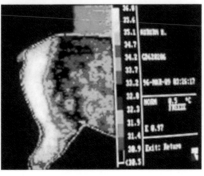

Figure 6: *A mare with pelvic mass impinging on lumbo-sacral plexus, causing right hind leg lameness. Purohit RC, Hanson R. Unpublished case, 2006.*

Thermography for the diagnosis of neurovascular injuries

The cutaneous circulation is under sympathetic vasomotor control. Peripheral nerve injuries and nerve compression can result in skin surface vascular changes that can be detected thermographically. Inflammation and nerve irritation may result in vasoconstriction, causing cooler thermograms in the afflicted areas. In contrast, transection of a nerve, and/or nerve damage to the extent that there is a loss of nerve conduction, results in a loss in sympathetic tone, which causes vasodilatation indicated by an increase in the thermogram temperature. This simple rationale is more complicated with different types of nerve injuries (neuropraxia, axonotomesis, and neurotmesis). Furthermore, lack of characterization of the extent and duration of injuries may make thermographic interpretation difficult.

Studies performed with various animal species show that if thermographic examination is performed properly under controlled conditions, it can provide an accurate diagnosis of neurovascular injuries. The following study examples include Horner's Syndrome, long term effects of neurectomies, and differential diagnosis of vascular versus nerve injuries.

Horner's Syndrome

Clinical cases, along with surgical induction of Horner's Syndrome by transection of the vagosympathetic trunk in horses, have been reported by Purohit and McCoy.[17] Facial thermograms of Horner's Syndrome were done 15 minutes before and after the exercise. The affected side was warmer by 2–3 °C more than the non-transected side (Figure 7). This increased

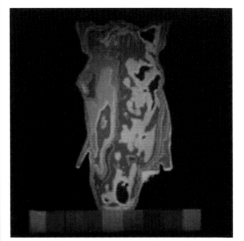
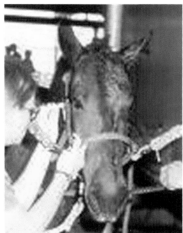

Figure 7: *Left Horner's Syndrome*

temperature after denervation is reflective of an increase in blood flow due to vasodilatation in the denervated areas.[17,22] The increased thermal patterns on the affected side were present for up to 6–12 weeks. In about 2–4 months, blood flow on the neurotraumatized side readjusted to the local demand of circulation. Thermography of both non-neurectomized and neurectomized sides looked similar and normal and it was difficult to distinguish the affected side. The intravenous injection of 1 mg of epinephrine in a 1,000 lb. horse caused an increase in thermal patterns on the denervated side, indicating the presence of Horner's Syndrome. Administration of I.V. acetylpromazine (30 mg. 1,000 lb. horse) showed increased heat (thermal pattern) on the normal, non-neurectomized side, whereas acetylpromazine had no effect on the neurectomized side. The alpha-blocking drug acetylpromazine caused vasodilation and increased blood flow to the normal, non-neurectomized side, whereas no effect was seen in the neurectomized side due to the lack of sympathetic innervation.[17,19,22]

Effects of Neurectomies

Thermographic studies were performed on the thoracic (front) and pelvic (back) limbs of horses before and after neurectomies. After posterior digital nerve neurectomy, there were significant increases in heat in the areas supplied by the nerves.[22] Within 3–6 weeks, readjustment of local blood flow occurred in the neurectomized areas, and it was difficult to differentiate between the non-neurectomized and the neurectomized areas. Ten minutes after administration of 0.06 mg/kg I.V. injection of acetylpromazine, a 2–3 °C increase in heat was noted in normal non-neurectomized areas, whereas the neurectomized areas of the opposite limb were not affected.

Vascular Injuries

It has been shown that localized reduction of blood flow occurs in horses with navicular disease.[29] This effect was more obvious on thermograms obtained after exercise than before exercise. Normally, 15–20 minutes of exercise will increase skin surface temperature by 2–2.5 °C in horses.[13,29] In cases of arterial occlusion, the area distal to the occlusion in the horses' limbs shows cooler thermograms. Exercise or administration of alpha-blocking drugs such as acetylpromazine causes increased blood flow to peripheral circulation in normal areas with intact vascular and sympathetic responses.[19,22] Thus, obtaining thermograms either after exercise or after administration of alpha-blocking drugs such as acetylpromazine provides prognostic value for diagnosis of adequate collateral circulation. The use of skin temperature as a measure of skin perfusion merits consideration for peripheral vascular flow and perfusion, despite some physical and physiological limitations, which are inherent in methodology.

Neurogenic inhibition can be diagnosed through the administration of alpha-blocking drugs, which provide an increase in blood flow. Vascular impairment may also be associated with local injuries (inflammation, edema, swelling, etc.), which normally provide localized cooler areas on thermograms.

Musculoskeletal Injuries

In clinical and subclinical cases of osteoarthritis, tendonitis, navicular disease, and other injuries such as sprains, stress fractures, and shin splints,[13,29–31] thermography has been efficacious for the diagnosis of these conditions. In some cases, thermal abnormalities may be detected two weeks prior to the onset of clinical signs of lameness in horses, especially in the case of joint disease,[31] tendonitis,[13] and navicular problems.[29,30]

Osteoarthritis is a severe joint disease in horses. Normally diagnosis is made by clinical examination and radiographic evaluation. Radiography detects the problem after deterioration of the joint surface has taken place. Clinical evaluation is only performed when horses show physical abnormalities in their gait due to pain. An early sign of osteoarthritis is inflammation, which can be detected by thermography prior to becoming obvious on radiograms.[31]

Thermography was used to evaluate the efficacy of corticosteroid therapy in amphotericin-B induced arthritis in ponies.[2] The chronic and acute pain associated with neuromuscular conditions can also be diagnosed by this technique. In cases where no definitive diagnosis can be made using physical examination and x-rays, thermography has been efficacious for early diagnosis of soft tissue injuries.[13,23]

Studies of Thermoregulation of the Testes and Scrotum

The testes of most domestic mammalian species migrate out of the abdomen and are retained in the scrotum, which provides the appropriate thermal environment for normal spermatogenesis.[32,33] The testicular arterial and venous structure is such that arterial coils are enmeshed in the pampiniform plexus of the testicular vein, which provides a countercurrent heat-regulating mechanism by which arterial blood entering the testes is cooled by the venous blood leaving the testes.[32,33] Thus, to function effectively, the mammalian testes are maintained at a lower temperature. Purohit[15,16,18] used thermography to establish normal thermal patterns and gradients of the scrotum in bulls, stallions, bucks, dogs, and llamas. The normal thermal patterns of the scrotum in all species studied are characterized by right-to-left symmetrical patterns, with a constant decrease in the thermal gradients from the base to the apex. In bulls, bucks, and stallions, a thermal gradient of 4–6 °C from the base to apex with concentric bands signifies normal patterns. Inflammation of one testicle increased ipsilateral scrotal temperatures of 2.5–3 °C.[16,18,35] If both testes were inflamed, there was an overall increase of 2.5–3 °C temperature and a reduction in temperature gradient was noted.

Testicular degeneration could be acute or chronic. In chronic testicular degeneration with fibrosis, there was a loss of temperature gradient, loss of concentric thermal patterns, and some areas were cooler than others with no consistent patterns.[16] The infrared thermal gradients and patterns in dogs[18] and llamas[8,18] are unique to their own species and the patterns are different from that of the bull and buck. In humans, normal thermal patterns are characterized by symmetric and constant temperatures between 32.5 and 34.5 °C.[1,4,6,9,10,34] Increased scrotal infrared emissions were associated with intrascrotal tumor, acute and chronic inflammation, and varicoceles.[4,6] Thermography has been efficacious for early diagnosis of acute and/or chronic testicular degeneration in humans and many animal species.

Conclusions

All species studied thus far have provided remarkable bilateral symmetrical patterns of infrared emission. The high degree of right-to-left symmetry is a valuable asset in diagnosis of unilateral problems associated with various inflammatory disorders. On the other hand, bilateral problems can be diagnosed due to changes in thermal gradient and/or overall increase or decrease of temperature, away from the normal, established thermal patterns in a given area of the body. Various areas of the body on the same side have normal patterns and gradients. This can be used to diagnose a change in gradient patterns. Alteration in normal thermal patterns and gradients indicates a thermal pathology. If thermal abnormalities are evaluated carefully, early diagnosis can be made, even prior to the appearance of clinical signs of joint disease, tendonitis, and various musculoskeletal problems in various animal species. Thermography can be used as a screening device for early detection of an impending problem, allowing treatment before the problem becomes more serious.

The efficacy of non-contact, electronic infrared thermography has been demonstrated in numerous clinical settings and research studies as a diagnostic tool for veterinary medicine.

References

1. Amiel JP, Vignalou L, Tricoire J, Jamain B, and Ravina JH. Thermography of the testicle: preliminary study. *J Gynecol Obstet Biol Reprod*, 1976;5:917–23.

2. Bowman KF, Purohit RC, Ganjam VK, Pechman RD Jr, and Vaughan JT. Thermographic evaluation of corticosteroid efficacy in amphotericin-B induced arthritis in ponies. *Am J Vet Res*, 1983;44:51–6.

3. Clark JA and Cena K. The potential of infrared thermography in veterinary diagnosis. *Vet Rec*, 1977;100:404.

4. Comhaire F, Monteyne R, and Kunnen M. The value of scrotal thermography as compared with selective retrograde venography of the internal spermatic vein for the diagnosis of "subclinical" varicocele. *Fertil Steril*, 1976;27:694–8.

5. Delahanty DD and Georgi JR. Thermography in equine medicine. *J Am Vet Med Assoc*, 1965;147:235–8.

6. Gold RH, Ehrlich RM, Samuels B, Dowdy A, and Young RT. Scrotal thermography. *Radiology*, 1977;122:129–32.

7. Heath AM, Navarre CB, Simpkins A, Purohit RC, and Pugh DG. A comparison of surface and rectal temperature between sheared and non-sheared alpacas (Lama pacos). *Small Ruminant Res*, 2001;39:19–23.

8. Heath AM, Pugh DG, Sartin EA, Navarre B, and Purohit RC. Evaluation of the safety and efficacy of testicular biopsies in llamas. *Theriogenology*, 2002;58:1125–30.

9. Lazarus BA and Zorgniotti AW. Thermo-regulation of the human testes. *Fertil Steril*, 1975;26:757–9.

10. Lee YT and Gold RH. Localization of occult testicular tumor with scrotal thermography. *JAMA*, 1976;236(17):1975–6.

11. Nelson HA and Osheim DL. Soring in Tennessee walking horses: detection by thermography. *USDA-APHIS, Veterinary Services Laboratories*, Ames, Iowa, 1975, 1–14.

12. Purohit RC and Franco BD. Infrared thermography for the determination of cervical dermatome patterns in the horse. *Biomed Thermol*, 1995;15:213.

13. Purohit RC and McCoy MD. Thermography in the diagnosis of inflammatory processes in the horse. *Am J Vet Res*, 1980;41:1167–74.

14. Purohit RC, Bergfeld WA III, McCoy MD, Thompson WM, and Sharman RS. Value of clinical thermography in veterinary medicine. *Auburn Vet*, 1977;33:104–8.

15. Purohit RC, Carson RL, et al. Peripheral neurogenic thermoregulation of the bovine scrotum. *Thermology Int*, 2007;17(4):137–9.

16. Purohit RC, Hudson RS, Riddell MG, Carson RL, Wolfe DF, and Walker DF. Thermography of the bovine scrotum. *Am J Vet Res*, 1985;46,2388–92.

17. Purohit RC, McCoy MD, and Bergfeld WA 3rd. Thermographic diagnosis of Horner's syndrome in the horse. *Am J Vet Res*, 1980;41:1180–2.

18. Purohit RC, Pascoe DD, Heath AM, Pugh DG, Carson RL, Riddell MG, and Wolfe DF. Thermography: its role in functional evaluation of mammalian testes and scrotum. *Thermol Int*, 2002;12:125–130.

19. Purohit RC, Pascoe DD, Schumacher J, Williams A, and Humburg JH. Effects of medication on the normal thermal patterns in horses. *Thermol Osterr*, 1996;6:108.

20. Purohit RC, Schumacher J, Molloy JM, Smith, and Pascoe DD. Elucidation of thoracic and lumbosacral dermatomal patterns in the horse. *Thermol Int*, 2003;13:79.

21. Purohit RC. History and research review of thermography in veterinary medicine at Auburn University. *Thermol Int*, 2007;17(4):127–32.

22. Purohit RC. The diagnostic value of thermography in equine medicine. *Proc Am Assoc Equine Pract*, 1980;26:317–26.

23. Purohit RC. Use of thermography in the diagnosis of lameness. *Auburn Vet*, 1987;43:4.

24. Smith WM. Applications of thermography in veterinary medicine. *Ann NY Acad Sci*, 1964;121:248–54.

25. Stromberg B and Norberg I. Infrared emission and Xe-disappearance rate studies in the horse. *Equine Vet J*, 1971;1:7–14.

26. Stromberg B. The normal and diseased flexor tendon in racehorses. *Acta Radiol [Suppl]*, 1971;305:1–94.

27. Stromberg B. The use of thermograph in equine orthopedics. *J Am Vet Radiol Soc*, 1974;15:94.

28. Stromberg B. Thermography of the superficial flexor tendon in race horses. *Acta Radiol [Suppl]*, 1972;319:295–7.

29. Turner TA, Fessler JF, Lamp M, Pearce JA, and Geddes LA. Thermographic evaluation of podotrochlosis in horses. *Am J Vet Res*, 1983;44:535–9.

30. Turner TA, Purohit RC, and Fessler JF. Thermography: a review in equine medicine. *Comp Cont Educations Pract Vet*, 1986;8:854.

31. Vaden MF, Purohit RC, McCoy MD, and Vaughan JT. Thermography: a technique for subclinical diagnosis of osteoarthritis. *Am J Vet Res*, 1980;41:1175–9.

32. Waites GMH and Setchell BP. "Physiology of testes, epididymis, and scrotum," in *Advances in Reproductive Physiology*, A. McLaren, Ed., Logos Press, London, Vol 4, 1969, 1–21.

33. Waites GMH. "Temperature regulation and the testes," in *The Testis*, A.D. Johnson, W.R. Gomes, and N.L. VanDemark, Eds., Academy Press, Inc., New York, Vol 1, 1970, 237–41.

34. Wegner G and Weissbach Z. Application of plate thermography in the diagnosis of scrotal disease. *MMW*, 1978;120:61–4.

35. Wolfe DF, Hudson RS, Carson RL, and Purohit RC. Effect on unilateral orchiectomy on semen quality in bulls. *J Am Vet Med Assoc*, 1985;186:1291–3.

14

Psychosocial Implications of Thermography

EDWIN F. RICHTER III, MD

During their training, contemporary clinicians may often be advised of the importance of utilizing a biopsychosocial model to approach their patients' problems. This philosophy places significant emphasis on an individual's emotional well being. Attention is also paid to how that person interacts with family, friends, caregivers, employers, and other significant people in their social setting. The expectation is that ultimately this approach will better serve the patient's needs by going beyond a narrow biological approach to illness.

Sensitivity to psychosocial issues, however, should not diminish the clinician's appreciation of biological data. Any potentially useful information about underlying medical problems must be employed to refine the care plan. Objective facts, even if they may not immediately be used to drive decisions about medications or procedures, are fundamentally useful for education and counseling. They provide a framework on which abstract ideas can be based.

Profile of the chronic pain patient

These concepts should be remembered when considering the challenges presented by patients with chronic pain. This population of patients often seeks attention from multiple medical specialists. These consultations often result in conflicting diagnoses and conflicting advice. As time passes and more visits are experienced, a greater degree of uncertainty about their condition is experienced. Persistence of symptoms frustrates not only patients and their physicians but also family, friends, coworkers, and employers. This frustration is then exacerbated by the almost universal experience of the chronic pain patient, which is that others around them express doubt about the validity of their complaints.[8] This perception may develop into a problem that disturbs them as much as the pain itself.

People with chronic pain do not fail to perceive the doubts held by those around them. Relatives may express frustration over their failure to actively maintain certain social roles. Physicians, therapists, and other caregivers may fail to conceal their dismay when the chronic pain patient returns to the office or clinic. Employers and/or coworkers may express resentment about continued absence or requests for restricted duty. Awareness of these reactions may encourage pain behaviors, which are implicit external signals of the symptoms felt internally. There is a concern that the magnitude of these behaviors may increase to pursue ongoing sympathy and support. This can be perceived by clinicians and insurance carriers as evidence of malin-

> Persistence of symptoms frustrates not only patients and their physicians but also family, friends, coworkers, and employers.

gering in pursuit of financial compensation. Alternatively, such behavior may be interpreted as an attempt to get increased pain medication. This is particularly sensitive when patients with chronic nonmalignant pain are taking narcotics or other scheduled medications.

Some individuals are regarded with greater suspicion, because of aspects of their history. Any suggestion of a premorbid history of psychiatric illness may be seized upon as a reason to discount current complaints. Similarly, prior histories of substance abuse may diminish patients' credibility in some evaluators' opinions. Complaints of pain may be discounted if the patient is suspected of seeking prescriptions for controlled substances. Conversely, if a patient recovering from substance abuse tries to avoid use of such drugs, then the severity of their pain may be questioned.

Patients with nonmalignant pain may encounter a wide range of attitudes among clinicians about medication approaches. Some physicians are reluctant to prescribe narcotics in such cases, despite the recommendations of pain management specialists who may be more comfortable with such an approach. The chronic pain patient who is getting a reasonable benefit from treatment with narcotics supervised by an experienced specialist may therefore be labeled as an addict by other physicians. Other patients, who do not show improvements in function or quality of life on narcotic regimens, may be continued on those medications to help "prove" that something is wrong with them, or to get them out of their doctors' offices.

There are often many "stakeholders" concerned with a case of chronic pain. Aside from the obvious examples of the patients and their doctors, there are the patient's significant others; any current, former or future caregivers; insurance carrier representatives; attorneys; employers, coworkers or business associates; and all of the regulatory agencies that have jurisdiction over any of these parties. There are potential conflicts between the patient and any of these other interested parties. Many of them have definite interests in minimizing or maximizing the extent of any injury and resulting disability that may have occurred.

The potential for subtle and overt bias to enter this situation must be recognized. Specific financial interests of various parties may of course lead them to view the subjective aspects of a case differently, but more subtle factors might also play a role. Cultural and educational factors influence how many individuals present their situations, and this might lead to negative responses on the part of people from different backgrounds. The potential for such bias does not require radical differences in the race or ethnicity of the parties involved. A patient who presents his story using poor grammar, a strong regional accent, or frequent profanity may make a negative impression on some listeners. This may lead them to then discount his reliability. Colorful or dramatic presentations may be considered evidence of hysteria. A relatively sophisticated account of a medical history may be considered a sign of a "professional patient," who has become overly involved in their condition.

There are a number of plausible reasons why the credibility of a patient's complaints may be evaluated quite differently by various parties. A patient who becomes aware of doubts about the presence or severity of his pain may become overtly angry and bitter. This development is not likely to facilitate good therapeutic relationships with caregivers or community supports. Demonstration of a hostile attitude toward personnel initially approaching the case with sympathetic attitudes will tend to discourage them. Overt hostility toward an independent examiner or agency official might quickly antagonize someone who would have tried to approach an evaluation in a fair and ethical manner.

Evaluation of objective data

Evaluators may interpret supposedly objective data in different ways when influenced by various factors. There are many diagnostic tests that require judgement calls. Many physical examination findings require close attention to subtle distinctions, and honest differences of opinion may arise under the best of circumstances. One's specialty, academic background, and clinical experience may influence the likelihood of noting a particular finding. Interpretation of radiographic findings often involves decisions about whether abnormalities are mild or severe. Electromyographic findings are also commonly presented with qualitative assessments of severity. When reports are reviewed, there is usually no background information about how the interpreter makes these determinations. Even if one could obtain benchmark data about how a given radiologist or electromyographer tended to evaluate studies compared to their peers, that would not indicate with certainty how accurately they had assessed any one particular case.

Quantitative data, such as nerve conduction velocities, bone density measurements, or erythrocyte sedimentation rates, may be described in different subjective terminology. A given numerical value may be described as "abnormal," "minimally abnormal," or "insignificantly abnormal" by three different physicians. Conceivably, each choice could be justified, since a minimal deviation from a standard value is certainly abnormal but may not have practical significance. The implications of these choices of words, however, can be considerable.

Clinical correlation is essential when interpreting test results.

Clinical correlation is essential when interpreting test results. If an MRI or CT scan demonstrates an anatomic abnormality, the clinician must not only consider the apparent severity and location of the structural problem. The radiologic findings must be evaluated for their potential to cause physiological dysfunction. Sometimes this is relatively easy. A small osteophyte on the anterior aspect of the spinal column several levels away from the suspected area of pathology may be fairly easy to dismiss. A disc herniation in the right vicinity is clearly more suspicious, but cases of individuals with asymptomatic herniations may be cited to complicate matters.

Conversely, normal or rather equivocal radiologic findings will not disprove the possibility that a patient is experiencing severe pain. A false negative study may be obtained when imaging slices just missed the site of pathology. Any pathological problems that are generally not detected on conventional radiologic imaging studies will probably not be found. An abnormality residing in a different anatomic region than the one studied will not be found. Failing to give contrast before computerized tomography or omitting the most appropriate types of magnetic resonance imaging may reduce the sensitivity of some studies.

We can apply these principles to a situation in which leg pain is the chief complaint. If a radicular component is suspected, then imaging the lumbar spine is appropriate. Any positive findings must be weighed for their ability to cause referred pain to the area in question. If there is reasonable doubt, then further studies to clarify the situation are in order. Failure to do so may lead to several undesirable situations. Serious pathology may go untreated, while the patient sinks further into despair. Incidental findings may be over-treated, exposing the patient to iatrogenic injury. Undue emphasis on minor or irrelevant findings may also reinforce the "sick" role, discouraging efforts to move forwards.

Negative findings must also be weighed carefully. Normal anatomy of the lumbar spine would not preclude development of a painful neuropathy or a complex regional pain syndrome. Premature efforts to dismiss patients' symptoms from these conditions by waving unremarkable X-ray reports at them will not help. Their frustration with the individual who does such a thing may spill over toward the next person who tries to help them. This may hasten labeling such patients as "difficult," erecting another barrier between them and effective care.

Aspects of treating chronic pain patients

Patients who are desperate for help are vulnerable in multiple ways. If their physicians are dismissing them as undesirable or unworthy, they may seek forms of treatment from practitioners of uncertain qualifications. They may lobby aggressively for invasive procedures in the absence of valid indications. Reasonable treatment options may be neglected if they feel that their complaints are not taken seriously enough by those who prescribe such treatments.

Their failure to get better may progressively impair the relationships between patients with chronic problems and their clinicians. Pragmatists working on a fee-for-service basis may see practical advantages when some patients come back repeatedly, but most practitioners are genuinely troubled by treatment failures. In the abstract, they may understand that medical science has its limitations at present. In practice, they may blame themselves or the patients for doing something wrong. Even if this is not overtly expressed, such sentiments may influence relationships.

The psychosocial implications of thermography (computerized infrared imaging or CII) must be considered in the context of the problems introduced above. Patients with chronic pain are at risk for being dismissed, for numerous reasons, as malingerers or hysterics. They may therefore develop adversarial or guarded relationships with medical professionals and other caregivers. Some may dramatize their problems seeking sympathy, worsening their image with providers. Others may give up, becoming withdrawn. Any modality that reduces the risk of these developments is therefore worthy of some attention.

Safety is an important issue when considering further diagnostic testing of the chronic pain patient. Even relatively low-risk procedures may have some complications, and repeating them increases the odds of some adverse occurrence. The principle of doing no harm may be set aside when frustration is mounting, prompting trials of potentially dangerous treatments on a relatively speculative basis. Patients are generally required to sign consent forms before undergoing invasive procedures. The problem is that desperate patients may not really be attending to any warnings about the potential risks and uncertain benefits of these interventions.

Cost is also a valid concern. If a particular test is reasonably likely to find the basis of an individual's severe chronic pain, then there are strong ethical arguments in favor of performing the test regardless of cost. If a test is very unlikely to prove helpful, then one may begin to argue that finite funds are available for health care expenses, and the concept of doing greater good for others with those funds can come into play. If a test has been performed several times without adding any value to the clinical management, then it becomes very difficult to support a request for funding another repetition. Given the vanishingly small odds that something will finally show up on the nineteenth or twentieth MRI of the same body part, the only anticipated benefit is that perhaps the patient will feel that his case is still being taken seriously. It may not be appropriate for a third party to fund this type of gesture instead of more plausibly effective activities. Desperate patients and families should not be encouraged to undergo financial hardship to undergo procedures at their own expense if there is no reasonable expectation of benefit.

The benefits of thermography for the chronic pain patient

Infrared imaging can help address many of the concerns that have been raised in this review. It is a safe, noninvasive modality. No exposure to radiation or chemicals is involved. Costs are relatively modest when compared to many other diagnostic procedures. As long as the data generated is interpreted accurately and applied appropriately, there is little if any downside exposure clinically for the patient.

This technique of measuring heat emission asymmetry is a means of documenting and measuring a particular physiologic abnormality. This therefore differs from tests that measure other physiologic parameters, such as nerve conduction velocity or spontaneous electrical activity in muscles. It also differs from tests of anatomic structures, such as X-rays or range of motion measurements. Since thermography is looking at a different process, there is at least a theoretical argument in favor of trying this test in situations where other approaches have left unanswered questions.

There is a body of literature supporting this concept. Hendler et al.[6] demonstrated that a significant number of patients previously dismissed as having psychogenic pain were found on thermographic testing to have objective evidence of an abnormality that correlated with their complaints. Rosenblum reported a series of patients with previously undiagnosed causalgia detected thermographically.[10] Similarly, a number of case reports and case series have been presented by members of the Kathryn Walter Stein Chronic Pain Laboratory of the Rusk Institute supporting the use of thermography in diagnosing chronic pain syndromes.[1–5,9,11]

> A significant number of patients previously dismissed as having psychogenic pain were found on thermographic testing to have objective evidence of an abnormality that correlated with their complaints.

A review of literature on reflex sympathetic dystrophy revealed the value of thermographic imaging. Studies that did not utilize this modality reported a much lower incidence of reflex sympathetic dystrophy than those that did.[7]

One critical aspect of thermography is that the results are quantitative and objective. Heat emission asymmetry is expressed in degrees. There may be some room for debate over how to distinguish between mild and moderate degenerative changes on a radiograph, but numerical data is readily understood and accepted. (With thermography, one is dealing with actual measured numbers, unlike judgement calls of 4+ versus 5- motor power, 1+ versus 2+ edema, and other somewhat arbitrary numerical assessments.) Competent interpreters should get essentially the same quantitative results when looking at a given set of images. Although various subtle biases may affect how the results are interpreted clinically, the numbers are themselves objective data. Another clinician can therefore read a report, note the presence and degree of asymmetry or other abnormal patterns, and start analyzing the results without wondering if one degree Centigrade means the same thing to all parties concerned. The same is not necessarily true when one reads of abnormal insertional activity or other descriptive terms used in electromyography.

The temperature data is independent of patient effort. In conventional testing, there is no need for the patient to lift any weight, move a joint, or do other tasks under voluntary control. Patient symptoms may be noted during testing, but the results of the test do not rely on whether the patient complains of pain, or how severe those complaints are. (Patients with paroxysmal pain alternating with absence of pain are a partial exception to this rule, as testing while no symptoms are present may miss relevant findings.) There is therefore no immediate need for an observer to look for signs of submaximal efforts or symptom magnification, as they do not influence the test results. This may be particularly useful in dealing with a patient who has developed some chronic pain behaviors on top of underlying pathology.

Computerized infrared images can be stored and copied like other types of electronic data. Raw test data can be interpreted by additional reviewers at any time after the procedure was done. This allows for "blind" interpretation by someone who was not present during the testing. This is often not the case with electrodiagnostic testing, in which only the examiner's interpretation of the data is available for future review.

Infrared imaging may reveal abnormal patterns of heat emission. These are physiologic abnormalities, as are findings of asymmetries of muscle bulk or reflex activity between left and right extremities. Such abnormalities are not diagnoses. They are simply findings, which can be correlated with the clinical history and other relevant findings. Although one may speculate about future use of artificial intelligence and sophisticated computerized algorithms to reach diagnoses, the current process requires interpretation by a human clinician.

Interpreting thermography results

Data from thermography will not necessarily correlate with the presence or absence of pain, nor will it prove or disprove the severity of pain. This is because temperature asymmetries are not the same thing as pain. Temperature abnormalities are frequently associated with painful areas, but the correlation is not exact. Physiological processes associated with painful conditions may also have direct or indirect effects on temperature regulation. In some cases, the infrared images correlate dramatically with the precise location of the painful areas. In other cases, the painful symptoms may be relatively localized, while the temperature changes are regional in nature. This may reflect differences in the organization of somatosensory data and autonomic functions in the nervous system. Caution should be used in avoiding promises that the test will show an exact map of the patient's pain.

A commonsense approach must be applied to many cases. If pain is reported in certain digits, but there is a profound temperature difference between the subject's hands across all digits, then one would likely accept that the test supported the likelihood of some pathologic process. If the thermographic findings are not localized to the distribution of a particular peripheral nerve or cervical root, then one should hesitate to diagnose a discrete nerve injury or radiculopathy on the basis of the test results. More fundamentally, however, one can indicate to the patient that there was an abnormal finding, which will be interpreted in the context of all the other relevant information.

We may then assume that in some cases an infrared study will be read as clearly abnormal, but no clear underlying diagnosis can be made. These studies may still have great practical value. If no other study showed any definite abnormality, then at least there is some objective data to consider. The information may stimulate further analysis of the case, and follow-up testing may be considered. Various trials of empiric therapy may be considered, since there is at least some evidence of an associated abnormal physiologic process.

Abnormal infrared results do not rule out possible concurrent psychosocial disorders. This reflects the fact that a patient with a somatization disorder may experience a legitimate physical injury. An individual with a history of substance abuse or dependence may seek analgesics for various reasons. There is some benefit in having objective data to consider when assessing such patients. There may be situations in which one can express awareness that there does appear to be something abnormal going on within a patient's body. With this being understood, then a discussion of a need for psychological counseling or psychopharmacologic management may be less readily interpreted as a dismissal of the patient's complaints. Patients may also understand that specific concerns about their requests for increased analgesics are not simply an expression of distrust.

At times, physicians may recommend that certain therapies should be discontinued or avoided. This may result from concerns about potential side effects, lack of evident benefit, or financial considerations. Chronic pain patients may be quite suspicious of such recommendations. Continuing to provide a treatment can be seen as proof that a clinician still believes the patient's complaints. If the clinician can discuss the merits of the actual treatment as an independent issue, then a consensus may be achieved. If both parties accept that an abnormal thermographic study did support the existence of an abnormal physiologic state, then there is one less factor encouraging the patient to pursue potentially useless or harmful interventions.

Clinicians may need to encourage patients with chronic pain to increase their activity level. Exercise may enhance physical conditioning and emotional well-being. Participation in social and recreational activities may also improve psychosocial function. A return to the workplace may enhance self-esteem. In these situations, where some escalation of pain is feared by the patient, the clinician must credibly explain the differences between "hurt" and "harm." This challenge can be better met if the patient accepts that a thorough evaluation has been done, and that the reality of the experience of pain is recognized.

Utilization of infrared technology will not eliminate the need for very careful counseling of the chronic pain patient. Caution must be used when presenting abnormal findings. One should avoid inflammatory language that would imply that prior diagnostic evaluations had been improper or inadequate. (Infrared testing is not universally available at this time, and is not routinely performed as a standard diagnostic procedure in many medical centers.) One also needs to consider that many of these patients are anxious and vulnerable. They may be inclined to interpret even casual remarks as indications of some terrible prognostic indicator.

Utilization of infrared technology will not eliminate the need for very careful counseling of the chronic pain patient.

The emotional state of chronic pain patients necessitates careful training of all staff members. Altered body images may discourage some patients from disrobing adequately for the study. Limited ability to maintain a given position may hinder ability to hold still for acclimation and imaging. Misplaced concerns about radiation exposure may exacerbate these problems if the technology was not clearly explained, in a manner consistent with the patient's educational background. These problems should not simply be dismissed as uncooperative behavior. The assistance of the referring clinician may be needed to work through such issues. The option of doing a more limited study may also be considered.

Patients may ask to see their images during or after testing. In such cases, they may be overly impressed by some relatively insignificant findings. This is particularly true when color images are presented. Computerized units allow many options when adjusting settings. One can choose the number of colors utilized, the temperature differences between adjacent elements of the color spectrum, and the range of temperatures that will be visible on the image. A few keystrokes or mouse clicks can therefore make the same body part bright red or dark blue. The naïve observer, conditioned to associate these colors with heat or cold, may leap to erroneous conclusions.

Although gray-scale imaging avoids the suggestions that color images convey, there are still pitfalls with this technique. The same choice about the range of temperatures that will show up as visible images on the screen must be made. A body part that is slightly warmer or cooler than the endpoints of this range will appear invisible on the screen. Such an image may be highly disturbing to some patients.

Subjects undergoing this procedure have little to do. Since they are expected to remain relatively immobile during acclimation and imaging, they might seek to strike up conversations with whoever is present. They may ask the examiner questions about their condition. Unless the testing is being done by a treating clinician, or with a treating clinician present, this can lead to some confusion. Giving any advice about a medical condition can be considered as establishing a doctor-patient relationship, with responsibilities and liabilities beyond performing and interpreting this particular test. Unless a physician performing or supervising the test plans to establish such a relationship, giving clinical advice is best left to the treating clinicians.

Some caution should also be considered when patients request immediate interpretations of studies. Technicians would do best to refer such requests to the physician who will do the official interpretation, since otherwise any perceived discrepancies between preliminary comments and the official report can cause great emotional upset. Some patients may then suspect that some ulterior motive prompted the apparent change of report. The real reason for the discrepancy may be based in actual differences in interpretation of the data. In other cases, the patient may have misunderstood the original remarks. Technical jargon can often confuse laypersons. The distinction between statistically significant and insignificant numerical differences may not be appreciated.

The concept of clinical correlation may not be readily understood by some patients. A small area of asymmetry may be present during imaging, but may not clearly relate to the clinical picture. Such complex issues may seem like routine matters to medical personnel, but not to the average person. It may not be possible to have a sophisticated discussion of such concepts in the few minutes while the patient is getting dressed and walking out the door. A treating clinician who knows the patient well has the best opportunity to discuss test results in depth.

There are several reasons why a treating clinician should take the lead role in reviewing test results with patients, family members, and other interested parties. Detailed knowledge of the case allows optimal clinical correlation. An established rapport with the patient enhances communication. A relatively coherent picture of the status, plans, and goals can be presented. Fewer mixed messages may result.

When a treating clinician is the primary interpreter of the thermographic images, then forming a coherent clinical picture is easiest. In cases where the referring physician is highly familiar with the procedure, then a well-written report should suffice to move the process forwards. (In complex cases, further review with the interpreter may be needed.)

At times, a referring physician may not be very familiar with the procedure. In those cases, proper communication is essential. The interpreter may need to explain the imaging techniques, underlying physiologic processes, and criteria for interpretation. A standard written report may not cover all these issues adequately, and some direct discussion may be required. Sending copies of images can be helpful, but one must remember the types of errors that an inexperienced observer might commit. Color images might often be over-read, and gray-scale images might be under-read. Quantitative analysis of measured temperature differences must be included in reports. These must be explained clearly to the referring clinician, and to other clinicians when appropriate.

Treating referring professionals courteously is always a desirable practice, but the main beneficiaries of good communication between clinical personnel are patients. The primary goal of thermography is to provide diagnostic information that will facilitate treatment. Enhancing the ability of patients and caregivers to understand the clinical situation is also a valid goal, since this can lead to certain psychosocial benefits. The quality of interactions between patients and other "stakeholders" may improve as a result of an improved common understanding of the situation.

Explaining studies to patients

A special effort must be made when patients view any imaging study. This is particularly true with thermography, where the images may have a relatively dramatic appearance. When color images are seen by patients, or whoever accompanies them, there is concern that they may try to formulate their own interpretation. The vivid colors on the screen may be deceptive. Any bright red areas, for example, tend to catch the eye. This may cause undue worry in a patient who does not understand the technology.

One might take care to explain that the assignment of colors by the computer software is somewhat arbitrary in nature. One could add that the appearance on the screen is influenced by several parameters under the control of the examiner. It would also be prudent to stress that interpretation of this test requires quantitative measurements, using the computer's ability to calculate temperature differences between relevant areas. This might reduce the emotional impact of viewing a striking set of images.

This touches on two larger areas of contention. The first is whether details of medical test results should be shared in depth with patients. One point of view might hold that most patients are not well qualified to interpret medical information. Modern standards have held that patients should be as fully informed as possible. The latter approach does place an obligation on the concerned clinician to do more than simply present data. Meaningful explanations must be provided, in a manner that will genuinely promote better understanding.

The second area of concern involves who should take the lead role in sharing this information with the patient, and with other concerned parties. This raises an issue of professional courtesy, since referring clinicians may wish to review test findings before the patient receives the information. Patients may understandably wish to get answers to their questions as soon as possible. Other clinicians, such as medical specialty consultants and allied health professionals, may be eager to receive the data as well. Insurance company personnel, attorneys, disability evaluators, and others may also have legitimate interests in this information.

Ethically, there is a priority on pursuing the best interests of the patient while avoiding doing harm. Given the complex nature of many of the cases referred for thermography, a team effort between the thermography staff and the referring clinician is likely to best meet the patient's needs. If the referral did not come from the primary care physician, or if multiple specialties are involved in the case, then of course the extent of the collaborative effort may need to be expanded.

Discretion is certainly appropriate in terms of how information should be communicated directly to the patient by thermography personnel. This is by no means based on paternalistic assumptions that patients should be kept in the dark about their medical conditions. The relevant issue is although that a skilled interpreter of infrared images may be able to use the technology well, the resulting information must be interpreted by someone who truly understands the individual case well. Communicating this concept to patients will certainly help their understanding of the process.

Given the complex nature of many of the cases referred for infrared imaging, a team effort between the thermography staff and the referring clinician is likely to best meet the patient's needs.

References

1. Cohen JM, Wu SSH, Cabrera IN, Haas F, and Lee MHM. The Physiological Documentation of Repetitive Strain Injury using Computerized Infrared Imaging-A Case Series. *Arch Phys Med Rehabil*, 2001;82:1498.

2. Cohen JM, Wu SSH, Yuhn SH, and Lee MHM. Computerized Infrared Imaging in the evaluation of chronic pain in patients in whom the standard diagnostic work-up is negative. *Am J Phys Med Rehabil*, 2003;82:245.

3. Cohen JM, Wu SSH, Yuhn SH, and Lee MHM. Computerized Infrared Imaging as an objective assessment tool in patients undergoing lumbar sympathetic blocks for Complex Regional Pain Syndrome-Type I. *Am J Phys Med Rehabil*, 2003;82:245.

4. Cohen JM, Wu SSH, Yuhn SH, and Lee MHM. Computerized Infrared Imaging as a Tool in Monitoring the Clinical Response to Acupuncture Treatment in a Patient with Chronic Abdominal Pain: A Case Report. *Arch Phys Med Rehabil*, 2003;84:A26.

5. Cohen, JM, Yuhn SH, and Lee MHM. The role of Computerized Infrared Imaging as an Objective Assessment tool in diagnosing Complex Regional Pain Syndrome and facilitating its treatment. *Arch Phys Med Rehabil*, 2004;85:E42.

6. Hendler N, Uematsu S, and Long D. Thermographic validation of physical complaints of 'psychogenic pain' patients. *Psychosomatics*, 1982;23:283-7.

7. Hooshmand, H. *Chronic Pain: Reflex Sympathetic Dystrophy Prevention and Management*. CRC Press, Boca Raton, Florida, 1993, p. 109.

8. Kleinman, A. *The Illness Narratives*. Basic Books, New York, 1988, p. 57.

9. Richter EF, Wu SSH, Cohen JM, Cabrera IN, and Lee MHM. Computerized Infrared Imaging as a Diagnostic Tool in Shoulder-Hand Syndrome. *Am J Phys Med Rehabil*, 2001;80:318.

10. Rosenblum JA. Documentation of Thermographic Objectivity in Pain Syndromes. Academy of Neuromuscular Thermography: *Clin Proc Postgrad Med*, 1986(March);59-61.

11. Wu SSH, Cohen JM, Richter E, Cabrera IN, and Lee MHM. Role of Infrared Imaging in the diagnosis of Complex Regional Pain Syndrome Type I in Post-CVA patients. *Arch Phys Med Rehabil*, 1999;80:1167.

Thermography and the Legal Field

JAY ROSENBLUM, MD
MARC LIEBESKIND, MD, JD

The Greek physician Hippocrates first suggested a connection between pain, disease, and "hot spots" on the body during the fifth century B.C.[1] Now, nearly 2,500 years later, modern medical science has resurrected Hippocrates' notion from the dustbin of history, developing it into an important tool for medical diagnosis and documentation. During the past three decades, thermography — a method for imaging the distribution of temperature over the body's surface — has been applied to the diagnosis and documentation of an extraordinarily diverse array of medical disorders

When thermography (Computerized Infrared Imaging or CII) first emerged on the medical scene in the late 1950s, its supporters advocated the method primarily as an adjunctive tool in the diagnosis of breast cancer. This application remains in use, although controversy rages on about is appropriateness. Over the years, published research has also suggested the utility of thermography in diagnosing, documenting, and monitoring the course of disorders including breast and skin cancers, male infertility, foot diseases, blood clots, leprosy, and many other findings.[4,8,10–16,19,21] In addition, several studies have illustrated the usefulness of thermography in making pre-surgery decisions, e.g., regarding selection of the optimum site for amputation of an (ischemic) limb.[20] Probably the most promising domain of application, however, concerns the diagnosis, documentation, and management of a wide range of pain syndromes.

Although technically imprecise, it is not too far from the mark to say that thermography provides a picture of pain. The subjective experience of pain is usually associated with changes in blood supply to the affected area of the body. Thermographic examination records temperature differences resulting from these alterations in blood flow. The thermogram

— literally a picture of heat — shows a multicolored pattern, where each color indicates a different temperature. A trained physician can interpret thermograms and thereby obtain information relevant to the existence, location, and sometimes, cause of a person's pain.

Thermography in the Medical and Legal Communities

The medical community's increasing reliance upon thermal imaging, particularly for the diagnosis and documentation of pain and soft tissue injuries, has led somewhat rapidly to medicolegal applications. Since the question of whether a plaintiff experiences pain lies at the crux of many lawsuits, it is not surprising that thermography has been introduced into the courtroom so soon after its introduction at the clinical level. Before the advent of thermography, both the medical and legal professions had been forced to rely solely upon their subjective judgments. Thermography offers, for the first time, the possibility of objective documentation (or refutation) of a plaintiff's complaints. For over 25 years thermography's rising acceptance in the medical and legal communities can be attested to by the early admission as evidence in New Jersey, Florida, Louisiana, and elsewhere.[6] However, introduction of thermography into the courtroom has also met with some criticism.

> Opponents of thermographic evidence argue that the technique is subject to several biases.

Opponents of thermographic evidence argue that the technique is subject to several biases. The first bias is that a patient involved in litigation has a strong motive to influence the outcome of the thermogram toward demonstration of pathology and pain. While experienced thermographers generally claim that such manipulation can be detected easily and eliminated by repeating the thermogram, skeptics remain unconvinced.

The second type of bias concerns the referral source. According to a large body of research on expectancy and experimenter effects, the results of a scientific test can be biased by expectations or desires of the experimenter who administers the test.[18] This influence need not be conscious or purposeful; it may be subtle and unconscious. Thus, physicians who administer thermograms in good faith may end up producing reports slanted in favor of the referral source. To this charge, advocates of thermography respond that the test is sufficiently objective to overcome any tendencies. Moreover, they claim that thermography is no more subject to such potential influences than other diagnostic tests.

The question of bias in medicolegal thermography

A study by Rosenblum looked at the question of bias in medicolegal thermography.[17] In this study, thermographic findings were analyzed by referral source and by whether or not a patient was included in litigation. The results of the research bear upon the use of thermography in court as well as the more general question of the technique's inherent objectivity.

In this study, an analysis of the records of 318 patients who underwent thermography in a Manhattan clinical practice was conducted. Thermography was employed only when deemed necessary in order to arrive at a conclusion with regard to a patient's complaints. No thermograms were administered solely for research purposes.

Readings of the thermograms were performed under conditions designed to approximate the typical clinical conditions when physicians produce reports. No effort was made to have another physician replicate and validate thermogram readings because this seldom happens before a physician submits a report; moreover, this would be problematic since a research design should attempt to maximize the detectability of bias if, in fact, it exists. Similarly, double-blind readings would have been counterproductive in view of the goals of the study. The study was designed to provide a reasonable chance for bias to emerge if, in reality, it did exist in typical cases. Each thermogram was evaluated as normal or abnormal. A notation was made as to whether or not a patient was involved in litigation and, for litigating patients, whether the referral source was defendant or plaintiff. The patient population for this study did not show any marked differences by referral source or litigation status with regard to results of the clinical exam or disability judgment.

Of 318 thermograms analyzed, 56% (178) were abnormal and 44% (140) were normal (Table 1 and Figure 1). The results were then analyzed in terms of whether or not the patient was involved in litigation. Of 262 patients involved in litigation, 56% (146) had abnormal thermograms and 44% (116) had normal ones. Of the 56 patients not involved in litigation, 57% (32) had abnormal thermograms and 43% (24) had normal thermograms. Thus, the distribution of normal and abnormal thermograms was roughly similar for patients involved in litigation and patients not involved in litigation.

For patients involved in litigation, thermograms results were broken down by source of the referral and categorized as either plaintiff or defendant. Of 216 defendant referrals, 56% (120) patients had abnormal thermograms and 44% (96) had normal thermograms. Of 46 plaintiff patients, 57% (26) had abnormal thermograms and 43% (20) had nor-

Table 1: Thermogram Results By Litigation Status

Litigation Status	Number of Patients	Percent Abnormal	Percent Normal
Involved in Litigation	262	55.7	44.3
Not Involved in Litigation	56	57.1	42.9
Total Patients Examined	318	56.0	44.0

Table 2: Thermogram Results By Referral Source

Referral Source	Number of Patients	Percent Abnormal	Percent Normal
Defendant	216	55.6	44.4
Plaintiff	46	56.5	43.5
Total Patients Involved	262	55.7	44.3

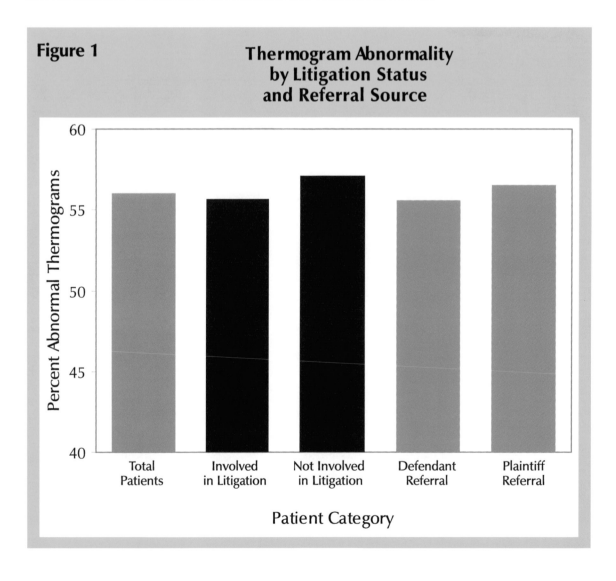

Figure 1

Thermogram Abnormality by Litigation Status and Referral Source

mal ones. Thus, referral source appears to have made very little difference in terms of the distribution of thermograms (Table 2 and Figure 1).

Thermography and the 'general acceptance' standard

Thermography has been successfully admitted into evidence in multiple jurisdictions. Appellate courts have ruled favorably on admitting infrared imaging technology into court to prove musculoskeletal injuries and as objective evidence of pain.[6]

Until the 1980s the 'general acceptance' standard traced to the Frye case held that scientific evidence could not be admitted in court unless it had "gained general acceptance in the particular field in which it belongs."[7] Applying this standard made admissibility difficult as the technique in question had to be shown to be a product of consensus before a jury could even hear the evidence. Over time, the Frye standard has relaxed considerably, most notably following the implementation of the Federal Rules of Evidence in 1975,[5] which state, "If scientific, technical, or other specialized knowledge will assist the prior effect to understand the evidence or to determine a fact in issue, a witness qualified as an expert by knowledge, skill, experience, training, or education, may testify thereto in the form of an opinion or otherwise."

Recent decades have seen conflict between the Federal Rules of Evidence and the Frye standard. The 'general acceptance' standard of Frye was increasingly found to be overly restrictive.[3] The conflict was resolved by the Supreme Court in Daubert v. Merrell Dow Pharmaceuticals, Inc., 509 U.S. 579 (1993). In its ruling, the Supreme Court bypassed the 'general acceptance' standard and emphasized the need for flexibility. Although they refused any "definitive checklist or test" for determining the reliability of expert scientific evidence, the Court did list several factors that it thought would be important:
  • whether the theories and techniques employed by the scientific expert have been tested
  • whether they have been subjected to peer review and publication
  • whether the techniques employed by the expert have a known error rate
  • whether they are subject to standards governing their application
  • whether the theories and techniques employed by the expert enjoy widespread acceptance

The emphasis in the ruling in Daubert v. Merell Dow Pharmaceuticals, Inc. was on whether the expert witness can assist the trier of fact. However, the floodgates were not simply thrown open to any and all testimony; testimony must be based on "scientific knowledge" in order to help establish evidentiary reliability. While little was stated to define this term, the testimony was to be limited by relevance or fit with the facts in dispute. The Court emphasized that the admissibility inquiry must focus "solely" on the expert's "principles and methodology," and "not on the conclusions that they generate." Moreover, the Daubert opinion underscored the traditional adversary system, noting cross examination, jury instruction, and burdens of proof.

Later cases have upheld the shift from the Frye standard to a flexible test in determining the admissibility of expert scientific testimony.[9] Factors to be considered include the specialized expertise of the witness and whether the testimony concerns research conducted independent of litigation.

In general, courts have an increasingly open-minded approach to new technologies as noted in one Pennsylvania case:[2]

"Medical technology is advancing more rapidly than the law. It is thus conceivable that a cutting-edge procedure, device or service may fall within the meaning of reasonable and necessary medical treatment, even though it has not gained general acceptance within members of the medical community."

References

1. Anbar M, Gratt BM, and Hong D. Thermology and facial tele-thermography. Part I: History and technical review. *Dentomaxillofac Radiol*, 1998 Mar;27(2):61-7.
2. Danko v. Erie Insurance Exchange, 630 A.2d 1219, 1222 (PA, Super. 1993), affirmed 538 Pa. 572. 649 A.2d 935 (1994).
3. Daubert v. Merrell Dow Pharmaceuticals, Inc., 509 U.S. 579 (1993).
4. Dribbon BS. Application and value of liquid crystal thermography. *J Am Podiatry Assoc*, 1983 Aug;73(8):400-4.
5. Federal Rules of Evidence, Rule 702 (1975).
6. Finnegan WJ and Koson DF. Jumping from the Frye Plan into the State Farm Fire: An Analysis of Spinal Thermography as Scientific Test Evidence. *Law, Medicine & Health Care*, 1985 Oct: 205-12.
7. Frye v. United States, 293 F. 1013 (D.C. Cir. 1923).
8. Jensen C, Knudsen LL, and Hegedus V. The role of contact thermography in the diagnosis of deep venous thrombosis. *Eur J Radiol*, 1983 May;3(2):99-102.
9. Kumho Tire Co., Ltd V. Carmichael, 526 U.S. 137 (1999).

10. Lewis RW and Harrison RM. Contact scrotal thermography II: use in the infertile male. *Fertil Steril,* 1980 Sep;34(3):259-63.

11. Lewis RW and Harrison RM. Contact scrotal thermography: application to problems of infertility. *J Urol,* 1979 Jul;122(1):40-2.

12. Libshitz HI. Thermography of the breast: current status and future expectations. *JAMA,* 1977;238:1953-4.

13. Manno E, Manno HD, Whitney D, and D'Amico JC. Thermographic application in podiatric medicine. *J Am Podiatry Assoc,* 1980 Apr;70(4):187-9.

14. Nyirjesy I, Abernathy MR, Billingsley FS, and Bruns P. Thermography and detection of breast carcinoma: a review and comments. *J Reprod Med,* 1977 Apr;18(4):165-75

15. Pochaczevsky R and Meyers PH. The value of vacuum contoured, liquid crystal, dynamic breast thermoangiography as an aid to mammography in the detection of breast cancer. *Clin Radiol,* 1979 Jul;30(4):405-11.

16. Ratz JL and Bailin PL. Liquid-crystal thermography in determining the lateral extent of basal-cell carcinoma. *J Dermatol Surg Oncol,* 1981 Jan;7(1):27-31.

17. Rosenblum JA. Documentation of Thermographic Objectivity in Pain Syndromes. Academy of Neuro-Muscular Thermography: Clin. Proc. Postgraduate Medicine. March 1986, pp.59-61.

18. Rosenthal R. *Experimenter Effects in Behavioral Research,* Appleton-Century-Crofts, New York, 1966.

19. Sabin TD. Temperature-linked sensory loss in leprosy. *Proc Am Thermographic Soc,* 1973:155-64.

20. Spence VA, Walker WF, Troup IM and Murdoch G. Amputation of the ischemic limb: selection of the optimum site by thermography. *Angiology,* 1981;32(3):155-69.

21. Wojeiechowski J and Zachrisson BF. Thermography as a screening method in the diagnosis of deep venous thrombosis of the leg. *Acta Radio Diagn* (Stockh), 1981;22(5):581-4.

Thermography and Art

Sandra H. Moon, MPH
Mathew H.M. Lee, MD, MPH

Introduction

While the use of thermal imaging has focused primarily on detecting pain and illness in human subjects as well as military and environmental uses, we have also been able to explore the aesthetic side of thermal imaging. This chapter is an amalgamation of thermographic art. It is a summation of laboratory observations on heat patterns generated by plants, animals, and inanimate objects, as well as a collection of professional thermographic art done in collaboration with Korean-born video artist Nam June Paik.

Thermography and Plants

Most plants take on their surrounding air temperature due to their slow metabolic systems. However, researchers have found that there are several hundred plant species that generate heat. Some of these plants include the dead-horse arum, eastern skunk cabbage, Asian sacred lotus, as well as the Amazon water lily. Images 1–3.

Image 1: *Thermography sunflowers I (orange)*

This is an infrared image of a vase full of sunflowers. It appears that the petals assumed the ambient room temperature more closely than the dense, seed-populated centers.

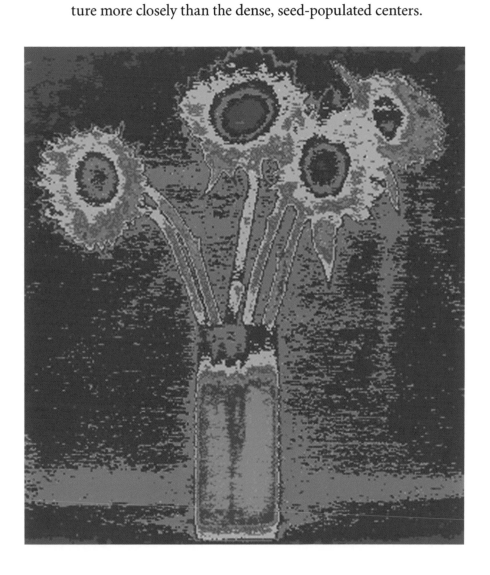

Image 2: *Thermography sunflowers II (purple/white)*

Image 3: *Thermography roses*

Thermography and Animals

Animals, like human subjects, emit heat from their bodies. However, unlike human subjects, thermographic images of animals vary according to the type and amount of skin, fur or feathers covering their bodies. Images 4–8.

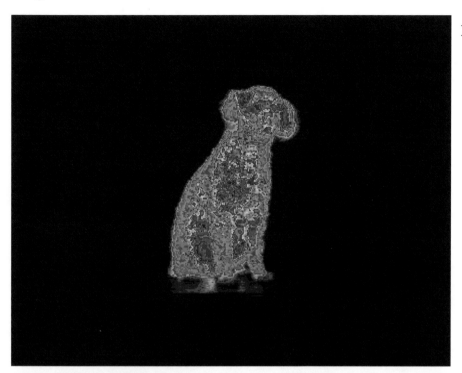

Image 4: *Dog sit*

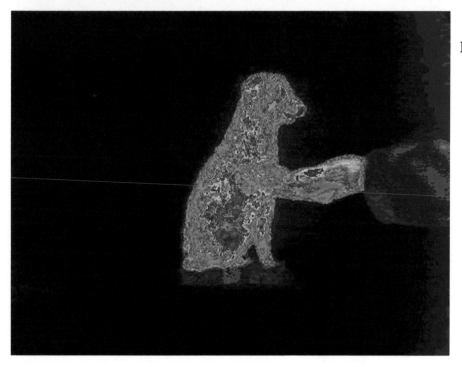

Image 5: *Dog shake*

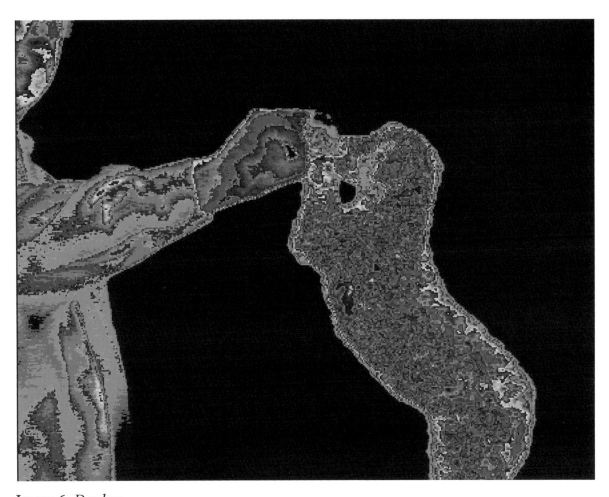

Image 6: *Dog beg*

Image 7: *Thermography cat*   **Image 8:** *Thermography bird*

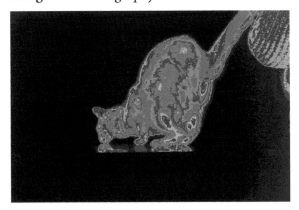   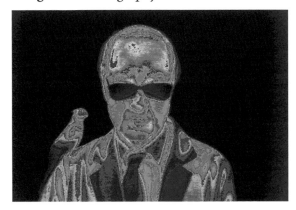

Thermography and Transfer of Heat on Inanimate Objects

According to Chinese philosophy, *chi* is the circulating flow of life energy inherent in all things. It is a belief that life energy does not simply disappear; it is either recycled or transferred. The following infrared images give a pictorial definition to this concept. Images 9–11.

Image 9: *Infrared image series of Dr. Lee's fingerprint on a plastic mold of his hand*

> This is a series of images capturing the energy transferred from a finger onto a clay sculpture of a hand. The white hot spot on the sculpture is evident immediately and a residual evidence of warmth remaining in the sculpture can be seen on the image many hours later.

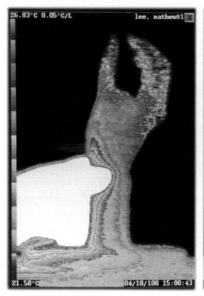 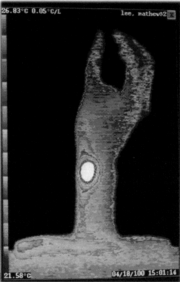 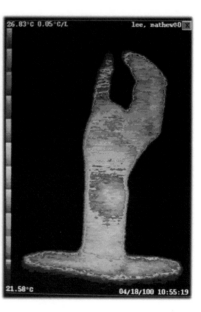

Image 10: *Infrared image series of football and handprint*

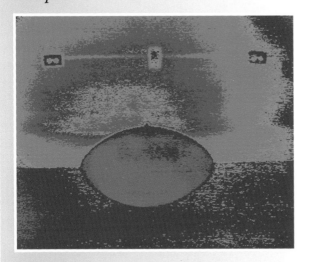

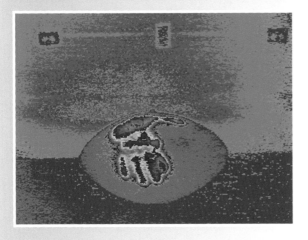

Image 11: *Infrared image series of cold iron, iron warming up (aura), hot iron*

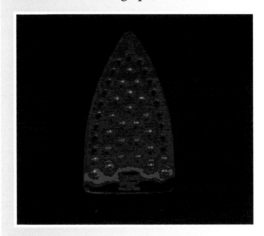

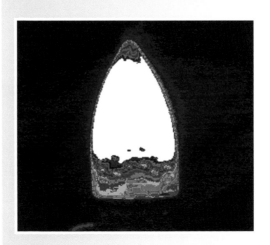

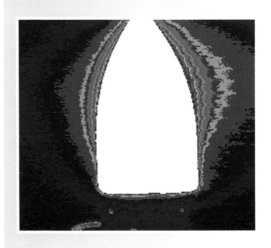

"Thermography: The Philosophy of Heat"

Nam June Paik is heralded as the father of video art, and his collaboration with thermography captures the beauty of the body's physiological changes as an expression of art. Images 12–15 were first displayed in an exhibition entitled "Thermography: The Philosophy of Heat" at the James Goodman Gallery in New York City on October 10, 2000 (see Appendix A for a review of the exhibition). These images not only portray natural heat emissions radiating from human subjects, but also the brilliant range of colors manifested by these emissions and creatively manipulated by the artist.

The thermography exhibit also traveled to the Marie Louise Trichet Gallery at the Wisdom House in Litchfield, Connecticut, in February of 2002 (see Appendix B for a review of this exhibit entitled "Thermographic Images"). Dr. Mathew Lee, of the Kathryn Walter Stein Chronic Pain Laboratory, presented a lecture on "The Artful Science of Pain" at the opening of the exhibit. Two pieces from this exhibit, "A Healer's Hand" and "The Artist," now reside in the permanent collection at the Whitney Museum of American Art in New York City.

These images and others were also included in a Cocktail Reception at the Union League Club in New York City on December 9, 2003 entitled "The Power, Beauty, and Vision of Thermography." The reception included a slideshow presentation, poetry reading, and appearances by James Goodman of the Goodman Gallery and artist Nam June Paik.

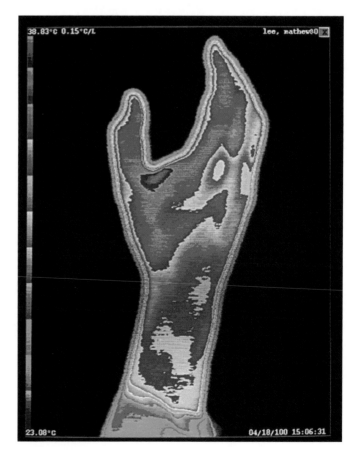

Image 12: *A Healer's Hand [ML]*
The hand of a physician who uses his hands to heal and comfort. The heat in his hand can represent the energy and sensitivity that is transmitted to his patients.

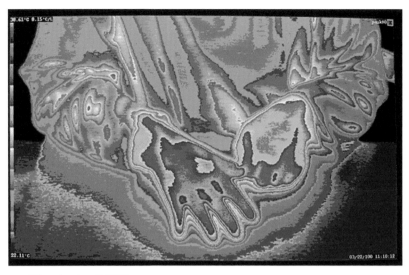

Image 13: *The Artist [NJP]*

> The hands of the gifted artist, Nam June Paik are filled with energy and life and this is captured in the intense warm colors of red and orange.

Image 14: *See You Later [NJP]*

> This infrared image taken of Nam June Paik was designed by the artist himself. By changing the temperature range recorded by the camera, the heat radiating from the hand and face are isolated, as if separated from the body.

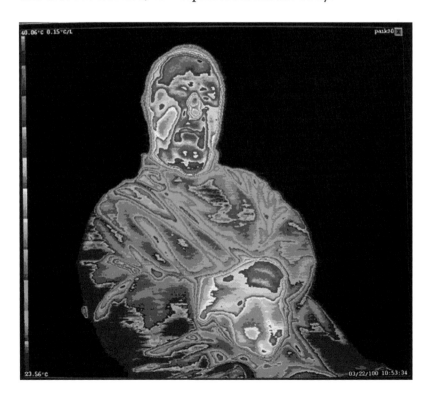

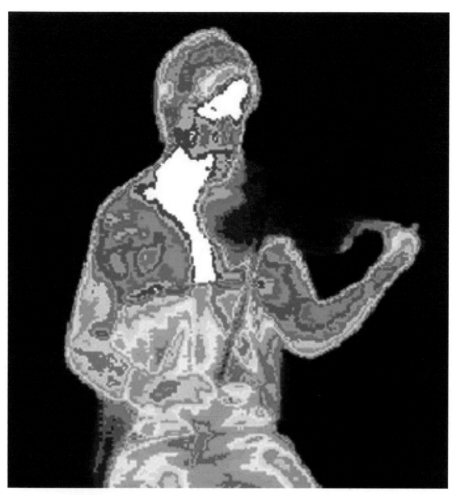

Image 15: A Violinst [YCM]
The infrared image of the violinist is taken while she is playing. We noticed that the violin remained surprisingly cool while the body of the violinist became warmer during playing.

Conclusion

The use of infrared images for the purposes of art captures the inherent beauty of animate and inanimate objects. The intensity and variety of colors, from red-hot to cool-blue, highlight the subtle points of energy contained throughout nature.

"The James Goodman Gallery was delighted to participate in a most unusual exhibition called *Thermography: The Philosophy of Heat,* billed as a collaboration between Dr. Mathew H.M. Lee and the Korean-born video artist Nam June Paik.

Robert C. Morgan, in his critique, wrote that he enjoyed the exhibition from the perspective of an aesthetics of science and the semi-abstract images, although it refers to the subject of pain, is part of the mystery of art.

The two images photographed in the Kathryn Walter Stein Laboratory are fascinating and reflect the beauty of nature. The Whitney Museum has accepted two of Dr. Lee's images in their permanent photographic collection."

James Goodman
James Goodman Gallery
Past President, Art Dealers' Association of America

Thermography:
A Medium for Conceptual Art?

BY ROBERT C. MORGAN
NY ARTS, NOVEMBER 2000 (VOL. 5 NO. 11)

Reprinted with permission of NY Arts and Robert C. Morgan, Ph.D.

The exhibition called "Thermography: The Philosophy of Heat" at the James Goodman Gallery is billed as a collaboration between Dr. Matthew H. M. Lee and the Korean-born video artist Nam June Paik. It is, in fact, the work of Dr. Lee using the head and body of Paik (among other subjects) as its subject matter. Thermography, as explained by Dr. Lee, one of the leading acupuncturists in the United States, is an instrument for measuring the radiation of heats from various parts of the body. The varying primary and secondary colors that reveal themselves in these extraordinary print-images are, in fact, coding devices in order to detect and measure the quality of heat emanating from various parts of the body. As an acupuncturist, Dr. Lee was interested in finding a way to measure the effect of his procedures on patients. The thermograph proved effective in this regard.

Apparently, Dr. Lee had seen Mr. Paik's retrospective at the Guggenheim Museum and became quite interested in the possibility of a collaboration. He was also interested in the potential of what he was doing with the thermographs as having some relationship to a new medium for art.

First of all, let me say that I have great respect for both the scientific work of Dr. Lee and for multimedia art endeavors of Nam June Paik, but the collaboration between art and science is always problematic. I recall years ago when Professor Gyorgy Kepes started the Center for Advanced Visual Studies at MIT, there was a similar aspiration — that somehow the future belonged to these kinds of collaborations. There was even the talk of dropping the word "art" in order that artists could become "researchers" and therefore discover or even reinvent what they were doing in relation to the new media. Certainly this interest persisted in some of the late work of the French postmodern theorist Jean-Francois Lyotard. At the risk of sounding conservative (which I am not), these kinds of collaborations are problematic. They are problematic because the paradigm of science will always rule over art.

The Dada artist Man Ray once wrote an essay in the twenties called "Photography is not art." What he meant by this was very obvious. It takes more than a new medium for art to happen because art is not just a matter of discovering a new medium. Art is not merely a technical process. On the art hand, we might say that art could become photography. This is true. But what makes this happen?

I think it happens because the artist has an intention, an idea, or a concept (a more evolved idea) that exceeds the technical apparatus. An intention is not merely an illustration of a technical apparatus or a systemic method. Neither is an artistic intention so academic as to predetermine all aspect of what the eventual work will be.

At the same time, there is no question that I enjoyed the exhibition; but I enjoyed from the perspective of an aesthetics of science. It was a happy exhibition, which sounds strange given that many of the subjects were suffering from pain of one kind or another when the thermographs were administered. Yet the paradox of seeing a semi-abstract image in a pleasurable [way], even [though] it refers to the subject of pain, is part of the mystery of art. Have you ever really looked at Tintoretto's Crucifixion in La Scuola San Rocco or an etching by Wols, an assemblage by Kienholz, or one of Artaud's self-portraits? These are works that depict pain, and perhaps there is a difference in these works from the thermographs taken at the [Kathryn Walter] Stein Chronic Pain Laboratory. There is a directness about the four works just named that expresses something that an illustration — in this case, a thermograph — cannot express. Science offers a necessary detachment in order to evaluate the evidence of what an image reveals — whether it is a chest X-ray, a sonogram, or a thermograph. Sometimes these documents reveal a special quality — an inexplicable visual, transsensory impact — that goes beyond their normative function in science. Given an artist's intention, these images can become art. Man Ray did this with his Rayograms, Rauschenberg did this with blueprints. The San Francisco Bruce Conner did this with his reappropriation of stock footage from films of the fifties depicting out-of-date experiments in psychology labs.

In the case of Dr. Lee, I think he is on to something, and I believe he should trust his insight. It takes more however than simply using the head and body or a famous artist as subject matter. Somehow these insights and intentions have to go beyond methodology of science in order to approach art.

Suffering for Art

*New Collaborative Exhibit at Wisdom House
Visually Displays Pain From Red Hot to Icy Blue*

BY JAIME FERRIS

*Reprinted with permission from Housatonic Publications,
January 25, 2002.*

So many people have experienced the phenomenon of pain, the internal response to stimuli that brings discomfort that can range from a tingle to the intolerable. Just speaking of such aches and throbs can make one cringe. But what if you could visually see pain, the vibrantly poetic colors it produces when viewed through the eye of an infrared camera?

That is exactly what visitors to the Marie Louise Trichet Art Gallery at Wisdom House Retreat and Conference Center in Litchfield can see in its newest exhibit, "Thermographic Images," which opens Feb. 2 with a special lecture by one of the artists. Dr. Mathew H.M. Lee, director of the Kathryn Walter Stein Chronic Pain Laboratory and the Howard A. Rusk Professor of Rehabilitation Medicine at the Rusk Institute of New York University School of Medicine, has worked in collaboration with renowned pioneer video artist Nam June Paik, to create an artistic "spin-off" of the medical process that illuminates the brilliance the body possesses. The infrared-camera images show the visionary work of thermography — bittersweet beauty in vibrant reds and oranges, warm yellows and cool greens and blues. About 19 photographs created through thermology, a method used in the scientific search for pain, are to be represented in the visually stunning display of images.

"There is just something about the images," said Marie Louise Trichet Art Gallery Director Jo-Ann Iannotti last week. "It is similar to outsider art, but there are so many connections on a physical and spiritual level and we want to share that with people. We feel so fortunate to have this exhibit here."

Ms. Iannotti explained that the exhibit resulted from attending the Caring Communities for the 21st Century conference at the United Nations last year. The basis for the conference, Ms. Iannotti said, was for people from all beliefs, religions, countries and cultures to gather and see what they can do to help unite communities. During the conference, Ms. Iannotti and Sister Rosemarie Greco heard an intriguing presentation by Dr. Lee called "Musical Rhythms, Technology and Health," describing the use of music during the process of rehabilitation.

After the presentation, the two women spoke to Dr. Lee in the lobby and, as the conversation progressed, he talked about his exhibit of thermographic images that were first unveiled to the public more than a year ago in the James Goodman Gallery in New York. He gave them a sample postcard and then corresponded with Ms. Iannotti, who brought the idea to the attention of the gallery committee.

"We have had very different exhibits, very unique. [The committee] discussed the possibility of holding the exhibit," she said, noting that one member immediately recognized Mr. Paik's name. "'Oh my,' she said. '[Nam June Paik] is the father of video [art]. We must do this.' It came through discussions that we could truly see the connections of spirituality," Ms. Iannotti added. "The art/religion connection has always been there, but we don't think about it, necessarily, when healing is needed."

"It is about combining the artistic and the scientific," said Izumi Cabrera, Dr. Lee's former research assistant, a New York University medical student and Dr. Lee's assistant on the Wisdom House exhibit and its layout. "It is similar to the work shown in New York City in that it shows the relationship to pain and temperature."

What is Thermology?

Ms. Cabrera explained that thermology measures the temperatures in the body, which is instrumental in finding the location of pain. The technology has also been used for breast cancer and other conditions, although Ms. Cabrera said it has not been definitively been proven to work for other ailments. According to Clinical Thermology Associates (CTA), an independent Diagnostic Center, thermology is a "diagnostic technique that measures physiological function by recording thermal heat emission," which can aid in indicating specific health problems using infrared rays that are invisible to the human eye.

Thermology is not a new technique. Historians trace its roots back to 400 B.C. when wet mud and clay was used to detect the point of an injury.

It was believed that wherever the diseased area was, the mud dried there first. History also quotes Hippocrates as stating that, "should one part of the body be colder or hotter than the other, disease is present in that part." Galileo created the first thermoscope sometime between 1592 and 1596, CTA reported, to decipher the relationships between body temperature and the state of health.

As the years passed technology advanced. Infrared light was used by the U.S. Army during World War II and the Korean War to keep track of troops movements. The first human thermogram was produced in 1954 and needed an hour-long exposure time. Just two years later Dr. Ray Lawson was the first doctor recorded to have used thermography in medical application.

Between the 1960s and the early 1980s, thousands of papers were written on the subject, with the United States, Japan and Europe serving as the primary researchers for the technology. But the term "clinical thermology" wasn't used until 1978 by Dr. George Chapman, who is credited with coining the phrase "in an effort to differentiate medical thermology as a stand-alone diagnostic test and the use of clinical findings in thermology to clarify a thermographic diagnosis and direct treatment course." As technology continued to advance, so did the use of thermography for medical use, which is now done digitally. Scientists are now working on a hand-held version of the infrared camera to make it easier to employ at the bedside.

"It is so beautiful," Dr. Lee said. "We can use it as a measure of discomfort, [by recording the heat found in the body], but prefer to use the term, 'philosophy of heat.'"

Seeing Pain For the First Time

According to Dr. Lee's literature, thermology measures physiological changes by recording the thermal heat emission from the body. In thermograms, the heat, or lack thereof, is displayed with colors, from black and blue to bright shades of orange, red and white. Shades and colors typically vary from one part of the body to the next, but in images of healthy bodies, such changes would be displayed in a symmetrical fashion. However, in cases where there is pain, an acute injury or disease present, these shades may be altered or varied in the image.

"Pain is relative and is different for each individual," Dr. Lee said. "People have different tolerance levels and experiences with pain. But with thermology, you don't have to ask. It's no longer feeling, but seeing."

"It is a complicated process," Izumi Cabrera, his assistant on the exhibit, explained. "It records the temperature of the area of focus. The only similarity to a regular camera is that it is unlike x-rays or CAT scans. There is no radiation, and it is non-invasive."

The instruments used are large, not easily portable and are connected to a computer, she explained, which displays the image of the area scanned. The procedure is so benign that patients could even be asleep, as long as they are lying in a position that allows the specific area to be seen. As for color correlations, Ms. Cabrera said the shades indicate different types of pain, whether it be in cold areas, depicted as black and blue, or in hot tones that range up through the spectrum to reds, yellows and white. Black and blue are considered the coldest tones, and as the area gets warmer, the colors change to purple, shades of red, yellow, orange and white, which is the hottest, or as Ms. Cabrera said, "off the scale." From this data, the doctor may calculate what is happening in the body.

"Wherever there are extremes, the body tries to balance it," she explained. "If there is an extreme, there is a physiological explanation for it."

Dr. Lee explained how pain endured over a six month period is defined as "chronic pain," which indicates a reduction in circulation and is depicted as "cool spots." These spots are blue or black in an image. Acute pain, such as a sprained wrist, is hot and depicted as shades of orange, yellow or white. He said the thesis in thermography is that these pains will appear in the picture, and that through analyzing the data and talking with the patient, thermography makes the difficult diagnosis and treatment process a little easier.

"Chronic pain has become a nemesis — you can't see a lesion," Dr. Lee said, noting the task is extremely difficult, but aids in ruling certain conditions out. "With acute pain, such as a sprained ankle, it is very hot, and you can see that. But the colors depend on the circulation in the body and [the type of injury]."

Thermology is also a tool that displays how methods of treatment for pain work. Ms. Cabrera told a story of a nurse who had pain in her hands, which were very cold and were detected in a thermographic image as blue and black. Dr. Lee, who is licensed in acupuncture, took a single acupuncture needle and stuck it into one of her hands, and then took an image over a period of five minutes and then again 15 minutes later. At that point, Ms. Cabrera said, the image had changed to "white hot." There have also been changes in colors after arthritis patients have taken their medication, Ms. Cabrera noted.

"So many people have disbeliefs [about] acupuncture or think it is all in someone's head, but [Dr. Lee] actually documented that there is a correlation," she said.

"It was very exciting," Dr. Lee said. "It was the first time it showed that [acupuncture] had a peripheral effect and relieves pain."

"Thermography," Mr. Paik said, "is a spin-off of the research, which measures the temperatures of various body parts. This measurement gives clues to our Chi points, which have shown mysteries of our life force since the time of Lao-tse's development of acupuncture."

'Thermographic Images'

In the medical world, Dr. Lee has become a player in the clinical thermographic field, as well as the field of acupuncture. He now spearheads what has been called "one of the largest, most innovative departments of the world" as he focuses his energies in "total rehabilitation needs of chronic, long-term disabled patients," according to his biographical account. He has worked largely in chronic disease, geriatric rehabilitation, the epidemiology of disability and chronic pain, authoring many books and contributing to publications and serving on editorial boards dealing with these topics.

Dr. Lee has also immersed himself in the field of music in rehabilitation and its role in healing. In his studies of chronic pain, he played an instrumental part in establishing the New York Society of Acupuncture and the American College of Acupuncture. But it is his work at the chronic pain laboratory at the Rusk Institute, where he created his scientifically artistic images, and his collaboration with Mr. Paik that Dr. Lee will discuss at the opening reception at Wisdom House Feb. 2.

According to Ms. Cabrera, Mr. Paik met the physician following the video artist's show at the Guggenheim Museum. Mr. Paik is known internationally as the father of video art, a movement that has spurred a new generation of artists to use the innovative medium. Once he established the use of video as a "viable artistic medium," Mr. Paik exhibited at many galleries and museums, including the Museum of Modern Art, the Smithsonian American Art Museum and the Guggenheim, sharing his video image presentations with the public.

But it wasn't until after Mr. Paik visited the Rusk Institute following a stroke that the duo would embark on a journey to create an exhibit displaying the relationship of pain and temperature.

"They were all from my laboratory," Dr. Lee explained. "They were never meant to be exhibited, but [Mr. Paik] saw the photos and said, 'Hey.' So I asked him if he would do a show with me, and he agreed. He is a brilliant artist in his field and a very good friend."

Dr. Lee said Mr. Paik's name is "like gold" and he is the top artist in his field. So once the show was ready, the galleries and studios in New York, Dr. Lee said, were "thrilled."

"In clinical thermology, each degree in surface temperature can be assigned a color," Dr. Lee said about the images displayed previously in New York. "Pain alters the image. Knowing Nam June Paik and witnessing his extraordinary show at the Guggenheim drew me to collaborate with him on this project — to capture the beauty of the body's physiological changes as an art expression."

The two men tackled a process of collecting images to create a medium that displays both the artistic and the scientific fields in which the men individually excel. Ms. Cabrera explained that the photographs in the original exhibit (19 of the original 30 are being displayed in Litchfield) are those that represent temperature variations in the heads, hands and feet of Mr. Paik, Dr. Lee, James Goodman and some of the patients Dr. Lee treats.

"Nam June Paik's [pieces] are more artistic — he is interested in the colors," Ms. Cabrera noted. "Then, the theme with Dr. Lee is more scientific, but he has always been interested in art."

The image of Mr. Paik's hands display deep reds and purples in his right hand, orange-red in the left, surrounded by hints of greens, blues and yellows in his wrists and arms. Another image depicts Dr. Yeou-Cheng Ma, a developmental pediatrician and violinist who works with Dr. Lee, as she plays her violin — the instrument a cool blue, her head and chest a variation of orange and red, while her arms change to a warm yellow.

"That is where the artistic connection is with Dr. Lee," said Ms. Cabrera. "I think working with him for as long as I have, it is an artistic expression that really takes [the viewer] on a scientific path."

As for producing the images, Ms. Cabrera said it is a long, technical process of getting the images, translating them digitally onto a computer screen, and then photographing them so the images can be translated to a bigger canvas. This, Ms. Cabrera said is a time-consuming process with great care being taken not to lose the translated image by making it so large that the pixels become pronounced.

Guests to the Marie Louise Trichet Art Gallery at Wisdom House Feb. 2 will be able to hear Dr. Lee discuss his work and the artwork he and his friend have created. There is the added treat that this will be the first time these images have been displayed in Connecticut.

"It's about the connection of pain and how the colors can be translated to images," Ms. Cabrera said. "It branches out from useful technology, but [one] that can be appreciated aesthetically as well."

"It's scientific, but so beautiful at the same time," Dr. Lee said. "They are gifts from my patients, and I want to share that with people. The images shouldn't go to waste, and people can appreciate them now. And a by-product [of exhibiting them], hopefully, is that people will learn about [thermology] and use it."

Index